ISBN 978-1-5280-2207-1
PIBN 10902186

English
Français
Deutsche
Italiano
Español
Português

www.forgottenbooks.com

Mythology Photography **Fiction**
Fishing Christianity **Art** Cooking
Essays Buddhism Freemasonry
Medicine **Biology** Music **Ancient**
Egypt Evolution Carpentry Physics
Dance Geology **Mathematics** Fitness
Shakespeare **Folklore** Yoga Marketing
Confidence Immortality Biographies
Poetry **Psychology** Witchcraft
Electronics Chemistry History **Law**
Accounting **Philosophy** Anthropology
Alchemy Drama Quantum Mechanics
Atheism Sexual Health **Ancient History**
Entrepreneurship Languages Sport
Paleontology Needlework Islam
Metaphysics Investment Archaeology
Parenting Statistics Criminology
Motivational

NBS SPECIAL PUBLICATION 559

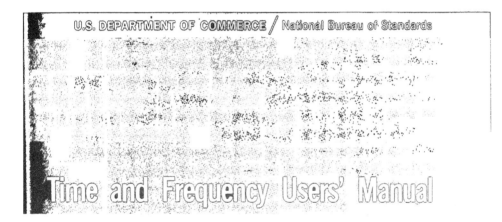

U.S. DEPARTMENT OF COMMERCE / National Bureau of Standards

Time and Frequency Users' Manual

NATIONAL BUREAU OF STANDARDS

The National Bureau of Standards[1] was established by an act of Congress on March 3, 1901. The Bureau's overall goal is to strengthen and advance the Nation's science and technology and facilitate their effective application for public benefit. To this end, the Bureau conducts research and provides: (1) a basis for the Nation's physical measurement system, (2) scientific and technological services for industry and government, (3) a technical basis for equity in trade, and (4) technical services to promote public safety. The Bureau's technical work is performed by the National Measurement Laboratory, the National Engineering Laboratory, and the Institute for Computer Sciences and Technology

THE NATIONAL MEASUREMENT LABORATORY provides the national system of physical and chemical and materials measurement; coordinates the system with measurement systems of other nations and furnishes essential services leading to accurate and uniform physical and chemical measurement throughout the Nation's scientific community, industry, and commerce; conducts materials research leading to improved methods of measurement, standards, and data on the properties of materials needed by industry, commerce, educational institutions, and Government; provides advisory and research services to other Government agencies; develops, produces, and distributes Standard Reference Materials; and provides calibration services. The Laboratory consists of the following centers:

Absolute Physical Quantities[2] — Radiation Research — Thermodynamics and Molecular Science — Analytical Chemistry — Materials Science.

THE NATIONAL ENGINEERING LABORATORY provides technology and technical services to the public and private sectors to address national needs and to solve national problems; conducts research in engineering and applied science in support of these efforts; builds and maintains competence in the necessary disciplines required to carry out this research and technical service; develops engineering data and measurement capabilities; provides engineering measurement traceability services; develops test methods and proposes engineering standards and code changes; develops and proposes new engineering practices; and develops and improves mechanisms to transfer results of its research to the ultimate user. The Laboratory consists of the following centers:

Applied Mathematics — Electronics and Electrical Engineering[2] — Mechanical Engineering and Process Technology[2] — Building Technology — Fire Research — Consumer Product Technology — Field Methods.

THE INSTITUTE FOR COMPUTER SCIENCES AND TECHNOLOGY conducts research and provides scientific and technical services to aid Federal agencies in the selection, acquisition, application, and use of computer technology to improve effectiveness and economy in Government operations in accordance with Public Law 89-306 (40 U.S.C. 759), relevant Executive Orders, and other directives; carries out this mission by managing the Federal Information Processing Standards Program, developing Federal ADP standards guidelines, and managing Federal participation in ADP voluntary standardization activities; provides scientific and technological advisory services and assistance to Federal agencies; and provides the technical foundation for computer-related policies of the Federal Government. The Institute consists of the following centers:

Programming Science and Technology — Computer Systems Engineering.

[1]Headquarters and Laboratories at Gaithersburg, MD, unless otherwise noted; mailing address Washington, DC 20234.
[2]Some divisions within the center are located at Boulder, CO 80303.

Time and Frequency Users' Manual

Edited by:

George Kamas
Sandra L. Howe

Time and Frequency Division
National Measurement Laboratory
National Bureau of Standards
Boulder, Colorado 80303

U.S. DEPARTMENT OF COMMERCE
Luther H. Hodges, Jr., Under Secretary
Jordan J. Baruch, Assistant Secretary for Science and Technology
NATIONAL BUREAU OF STANDARDS, Ernest Ambler, Director

Issued November 1979

Library of Congress Catalog Card Number: 79-600169

National Bureau of Standards Special Publication 559

Nat. Bur. Stand. (U.S.), Spec. Publ. 559, 256 pages (Nov. 1979)

CODEN: XNBSAV

(Supersedes NBS Technical Note 695, May 1977)

U.S. GOVERNMENT PRINTING OFFICE

WASHINGTON: 1979

For sale by the Superintendent of Documents, U.S. Government Printing Office.
Washington, D.C. 20402
Stock No. 003-003-02137-1 – Price $6.

(Add 25 percent additional for other than U.S. mailing).

CONTENTS

CHAPTER 4. USING TIME AND FREQUENCY IN THE LABORATORY

CHAPTER 6.   CALIBRATIONS USING LF AND VLF RADIO TRANSMISSIONS

CHAPTER 7. FREQUENCY CALIBRATIONS USING TELEVISION SIGNALS

CHAPTER 8. FREQUENCY AND TIME CALIBRATIONS USING TV LINE-10

CHAPTER 9. LORAN-C TIME AND FREQUENCY METHODS

CHAPTER 11. AN INTRODUCTION TO FREQUENCY SOURCES

CHAPTER 12. SUMMARY OF AVAILABLE SERVICES

xiii

ABSTRACT

This manual has been written for the person who needs information on making time and frequency measurements. It has been written at a level that will satisfy those with a casual interest as well as laboratory engineers and technicians who use time and frequency every day. It gives a brief history of time and frequency, discusses the roles of the National Bureau of Standards, the U. S. Naval Observatory, and the International Time Bureau, and explains how time and frequency are internationally coordinated. It also explains what time and frequency ser- vices are available and how to use them. It discusses the accuracies that can be achieved using the different services as well as the pros and cons of using various calibration methods.

Key Words: Frequency calibration; high fre- quency; Loran-C; low frequency; radio broad- casts; satellite broadcasts; standard fre- quencies; television color subcarrier; time and frequency calibration methods; time cali- bration; time signals.

PREFACE

This manual was written to assist users of time and frequency calibration services that are available in the U.S. and throughout the world. An attempt has been made to avoid complex derivations or mathematical analysis. Instead, simpler explanations have been given in the hope that more people will find the material useful.

Much of the information contained in this book has been made available in NBS Technical Notes. These were published in the last few years but have been edited and re- vised for this book. Many people and organi- zations have contributed material for this book, including other government agencies, international standards laboratories, and equipment manufacturers.

Since each topic could not be covered in great depth, readers are encouraged to request additional information from NBS or the respon- sible agency that operates the system or ser- vice of interest.

The OIL INDUSTRY needs accurate time to help automate oil well drilling, especially offshore.

JEWELERS & CLOCK/WATCH MANUFACTURERS need to set digital watches and clocks to the correct time before they leave the factory.

RAILROADS use time to set watches and clock systems. E.g., AMTRAK gets accurate time three times a day to set clocks in the AMTRAK system. This insures that trains arrive and depart on schedule.

The COMPUTER INDUSTRY needs accurate time for billing purposes, for timing the beginning and end of events for data processing, and for synchronizing communication between systems many miles apart.

POLICEMEN need time to check stopwatches used to clock speeders and they use frequency to calibrate radar "speed guns" used for traffic control.

SURVEYORS need time to measure distance and location. As already stated, 3 microseconds translates into 1 kilometer in distance when modern electronic instrumentation is used.

The COMMUNICATIONS INDUSTRY depends on accurate frequency control for its ability to deliver messages to its users. Time is needed for labeling the time of occurrence of important messages. For example, radio station WWV is used at the communications center at Yellowstone National Park.

The MUSIC INDUSTRY uses frequency (the 440-hertz tone from WWV, for example) to calibrate tuning forks which are used to tune pianos, organs, and other musical instruments.

MANUFACTURERS need time and frequency to calibrate counters, frequency meters, test equipment, and turn-on/turn-off timers in electric appliances.

The TRANSPORTATION INDUSTRY needs accurate time to synchronize clocks used in bus and other public transportation systems, and for vehicle location, dispatching, and control.

SPORTSMEN use time. Sports car rallies are timed to 1/100th of a second. Even carrier pigeon racers need accurate time.

The TELEPHONE INDUSTRY needs accurate time for billing purposes and telephone time-of-day services. Accurate frequency controls long-distance phone calls so that messages don't become garbled during transmission.

NATURALISTS studying wildlife habits want accurate time and frequency to help monitor animals they have fitted with radio transmitters.

The TELECOMMUNICATIONS INDUSTRY needs time accurate to one microsecond to synchronize satellite & other communications terminals spread over wide geographical areas.

ASTRONOMERS use time for observing astronomical events, such as lunar occultations and eclipses.

GEOPHYSICISTS/SEISMOLOGISTS studying lightning, earthquakes, weather, and other geophysical disturbances need time to enable them to obtain data synchronously and automatically over wide geographical areas. They use it for labeling geophysical events. Other SCIENTISTS use time for controlling the duration of physical and chemical processes.

Accurate time is required in MASTER CLOCK SYSTEMS in large institutions, such as airports, hospitals, large factories, and office buildings so that all clocks in the system read the same time.

The AVIATION/AEROSPACE INDUSTRY needs accurate time for aircraft traffic control systems and for synchronization at satellite and missile tracking stations. The FAA records accurate time on its audio tapes along with the air-to-ground communications from airplanes. Having an accurate record of when particular events happened can be an important factor in determining the cause of a plane crash or equipment malfunction.

MILITARY organizations use accurate time to synchronize clocks on aircraft, ships, submarines, and land vehicles. It is used to synchronize secure communications between command posts and outposts. Stable frequency will be necessary for navigation using a future satellite system.

- -

1.2 WHAT ARE TIME AND FREQUENCY?

We should pause a moment to consider what is meant by the word "time" as we commonly use it. Time of day or date is the most often used meaning, and even that is usually presented in a brief form of hours, minutes, and seconds, whereas a complete statement of the time of day would also include the day of the week, month, and year. It could also extend to units of time smaller than the second going down through milliseconds, microseconds, nanoseconds, and picoseconds.

We also use the word time when we mean the length of time between two events, called time intervals. The word time almost always needs additional terms to clarify its meaning; for instance, time of day or time interval.

Today time is based on the definition of a second. A second is a time interval and it is defined in terms of the cesium atom. This is explained in some detail in the chapter on atomic frequency sources (chapter 11). Let us say here that a second consists of counting 9,192,631,770 periods of the radiation associated with the cesium-133 atom.

The definition of frequency is also based on this definition. The term used to describe frequency is the hertz which is defined as one cycle per second.

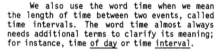

What does all this mean to users of time and frequency? Where does the laboratory scientist, industrial engineer, or for that matter, the man on the street go when he needs information about frequency and time measurements or about performing those measurements himself? That's what this book is about. It has been written for the person with a casual interest who wants to set his watch and for those with a specific need for frequency and time services to adjust oscillators or perform related scientific measurements.

This book has been deliberately written at a level which will satisfy all of these

TIME OF DAY:
9:00 A.M.

5 MINUTES
TIME INTERVAL

4

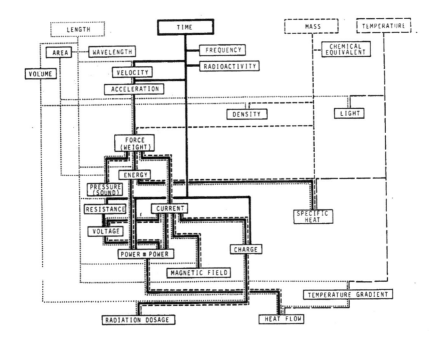

users. It will explain what is available in our world today in the form of services that provide time and frequency for many classes of users. In addition to providing information regarding these services, detailed explanations are given on how to use each service. Many countries throughout the world provide time and frequency services.

You can get time of day by listening to high frequency radio broadcasts, or you can call the time-of-day telephone services. You can decode time codes broadcast on various high, low, and very low frequency radio stations. You can measure frequency by accessing certain signals on radio and television broadcasts. You can obtain literature explaining the various services. And if your needs are critical, you can carry a portable clock to NBS or the USNO for comparison. This book describes many of these services and explains how to use them for time and frequency calibrations.

1.3 WHAT IS A STANDARD?

Of course, before you can have a standard frequency and time service, you have to have a standard, but what is it? The definition of a legal standard is, according to Webster, "something set up and established by authority, custom, or general consent as a model or example." A standard is the ultimate unit used for comparison. In the United States, the National Bureau of Standards (NBS) is legally responsible for maintaining and disseminating all of the standards of physical measurement.

There are four independent standards or base units of measurement. These are length, mass, time, and temperature. By calling them independent, we mean that all other measurements can be derived from them. It can be shown mathematically that voltage and pressure measurements can be obtained from measurements of these four base units. It is also true that frequency or its inverse, time interval, can be controlled and measured with the smallest percentage error of any physical quantity. Since a clock is simply a machine that counts frequency or time intervals, then time is kept with equal accuracy.

1.3.1 CAN TIME REALLY BE A STANDARD?

Time is not a 'standard" in the same sense as the meter stick or a standard set of weights. The real quantity involved here is that of time interval (the length of time between two events). You can make a time interval calibration by using the ticks or tones on WWV, for example, to obtain second, minute, or hour information, but you usually need one more piece of information to make that effort worthwhile. The information you need is the time of day. All national laboratories, the National Bureau of Standards among them, do keep the time of day; and even though it is not a "standard" in the usual sense, extreme care is exercised in the maintenance of the Bureau's clocks so that they will always agree to within a few microseconds with the clocks in other national laboratories and those of the U. S. Naval Observatory. Also, many manufacturing companies, universities, and independent laboratories find it convenient to keep accurate time at their facilities.

1.3.2 THE NBS STANDARDS OF TIME AND FREQUENCY

As explained later in this chapter and elsewhere in this book, the United States standards of frequency and time are part of a coordinated worldwide system. Almost the entire world uses the second as a standard unit of time, and any variation in time of day from country to country is extremely small.

But unlike the other standards, time is always changing; so can you really have a time standard? We often hear the term standard time used in conjunction with time zones. But is there a standard time kept by the National Bureau of Standards? Yes there is, but because of its changing nature, it doesn't have the same properties as the other physical standards, such as length and mass. Although NBS does operate a source of time, it is adjusted periodically to agree with clocks in other countries. In the next chapter we will attempt to explain the basis for making such changes and how they are managed and organized throughout the world.

So even though NBS does not have a glass-enclosed clock that is untouched or unadjusted, we do have a standard for frequency and time interval. This standard is located in Boulder, Colorado. It carries the designation NBS-6 because it is the sixth in a series of

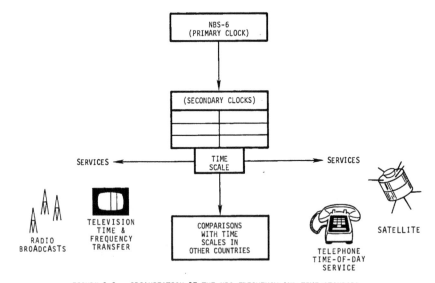

FIGURE 1.1. ORGANIZATION OF THE NBS FREQUENCY AND TIME STANDARD.

atomic oscillators built and maintained by NBS to provide the official reference for frequency and time interval in the United States. NBS-6 is referred to as the "master" or "primary" clock. It is used to calibrate other oscillators ("secondary" clocks) which are used to operate the time scale. (A time scale is a system of counting pendulum swings.) The organization of the NBS frequency and time standard is illustrated in figure 1.1. Shown is the primary atomic standard NBS-6, which is used to calibrate an ensemble of other atomic oscillators which operate the time scale, which is in turn compared and adjusted to similar time scales in other countries. Periodic adjustments are made to keep them all in agreement.

The useful output of the NBS standard and its associated time scale are the services provided to users. These services can take many forms and use several methods to get the actual needed calibration information to the user.

Notice that figure 1.1 looks a lot like any timepiece. We have a source of frequency, a means of counting that frequency and keeping time of day in a time scale, and as mentioned, a way of getting a useful output. This is exactly what is contained in almost every clock or watch in use today. The frequency source can be a 60 Hz power line, a balance wheel, tuning fork, or quartz crytal. The counter for totaling the cycles of frequency into seconds, minutes, and hours can be gears or electronics. The readout can be a face clock or digits.

In early times, the sun was the only timekeeping device available. It had certain disadvantages--cloudy days, for one. And also, you could not measure the sun's angle very accurately. People started developing clocks so they could have time indoors and at night. Surely, though, the sun was their "standard" instead of a clock. Outdoors, the sundial was their counter and readout. It is reasonable to suppose that if you had a large hour glass, you could hold it next to the sundial and write the hour on it, turn it over, and then transport it indoors to keep the time fairly accurately for the next hour. A similar situation exists today in that secondary clocks are brought to the master or primary clock for accurate setting and then used to keep time.

1.4 HOW TIME AND FREQUENCY STANDARDS ARE DISTRIBUTED

In addition to generating and distributing standard frequency and time interval, the National Bureau of Standards also broadcasts

time of day via its radio stations WWV, WWVH, and WWVB. Each of the fifty United States has a state standards laboratory which deals in many kinds of standards, including those of frequency and time interval. Few of these state labs get involved with time of day, which is usually left to the individual user, but many of them can calibrate frequency sources that are used to check timers such as those used in washing machines in laundromats, car washes, and parking meters.

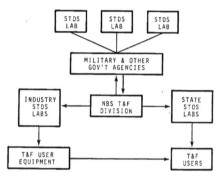

In private industry and other government agencies where standards labs are maintained, a considerable amount of time and frequency work is performed, depending on the end product of the company. As you might guess, electronics manufacturers who deal with counters, oscillators, and signal generators are very interested in having an available source of accurate frequency. This will very often take the form of an atomic oscillator kept in the company's standards lab and calibrated against NBS.

Although company practices differ, a typical arrangement might be for the company standards lab to distribute a stable frequency signal to the areas where engineers can use the signal for calibration. Alternatively, test equipment can periodically be routed to the standards lab for checking and adjustment.

As mentioned elsewhere in this book, the level of accuracy that can be achieved by a standards lab when making a frequency calibration ranges from a few parts per thousand to one part in a thousand billion, or 1 part in 10^{12}. Precise calibrations do involve more careful attention to details, but are not in fact that difficult to achieve.

1.5 THE NBS ROLE IN INSURING ACCURATE TIME AND FREQUENCY

The National Bureau of Standards is responsible for generating, maintaining, and

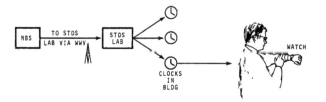

distributing the standards of time and frequency. It is not a regulatory agency so it does not enforce any legislation. That is, you cannot get a citation from NBS for having the incorrect time or frequency.

Other government agencies, however, do issue citations. Most notable among these is the Federal Communications Commission. This agency regulates all radio and television broadcasts, and has the authority to issue citations and/or fines to those stations that do not stay within their allocated frequencies. The FCC uses NBS time and frequency services to calibrate their instruments which, in turn, are used to check broadcast transmitters.

The Bureau's role is simply to provide access to the standards of frequency and time interval to users and enforcement agencies alike. When radio and television stations make calibrations referenced to NBS, they can be confident that they are on frequency and are not in violation of the broadcasting regulations.

1.6 WHEN DOES A MEASUREMENT BECOME A CALIBRATION?

Using an ordinary watch as an example, you can either move the hands to set the time of day, or you can change the rate or adjust the frequency at which it runs. Is this measurement a calibration? It depends on what reference you use to set the watch.

If the source you use for comparison is traceable at a suitable level of accuracy back to the National Bureau of Standards, (or the U.S. Naval Observatory in the case of DOD users), then you can say you have performed a calibration. It is very important to keep in mind that every calibration carries with it a measure of the accuracy with which the calibration was performed.

The idea of traceability is sometimes difficult to explain, but here is an example. Suppose you set your watch from a time signal in your laboratory that comes to you via a company-operated distribution system. If you

trace the signal backward toward its source, you will find that it perhaps goes through a distribution amplifier system to a frequency source on the manufacturing plant grounds. Let's say this source is an oscillator. Some means will have been provided to calibrate its output by using one of several methods--let's say NBS radio signals. With suitable records, taken and maintained at regular intervals, the oscillator (frequency source) can claim an accuracy of a certain level compared to NBS. This accuracy is transferred to your laboratory at perhaps a slightly reduced accuracy. Taking all these factors into account, the signal in your laboratory (and therefore your watch) is traceable to NBS at a certain level of accuracy.

What kind of accuracy is obtainable? As we said before, frequency and time can both be measured to very high accuracies with very great measurement resolution. As this book explains, there are many techniques available to perform calibrations. Your accuracy depends on which technique you choose and what errors you make in your measurements. Typically, frequency calibration accuracies range from parts per million by high frequency radio signals to parts per hundred billion for television or Loran-C methods. A great deal depends on how much effort you are willing to expend to get a good, accurate measurement.

1.7 TERMS USED

1.7.1 MEGA, MILLI, PARTS PER... AND PERCENTS

Throughout this book, we refer to suc things as kilohertz and Megahertz, millisec onds and microseconds. We further talk abou accuracies of parts in 10^9 or 0.5%. What d all of these terms mean? The following table explain the meanings and should serve as convenient reference for the reader.

Table 1.2 gives the meaning of 1 x 10- but what is 3 x 10-⁶? You can convert in th same way. Three parts per million or 3 x 10- is .0003%. Manufacturers often quote percen tage accuracies in their literature rathe than "parts per. . . ."

8

TABLE 1.1 PREFIX CONVERSION CHART

	PREFIX	DEFINITION	EXAMPLE
LESS THAN 1	MILLI	ONE THOUSANDTH	MILLISECOND (ms) = ONE THOUSANDTH OF A SECOND
	MICRO	ONE MILLIONTH	MICROSECOND (µs) = ONE MILLIONTH OF A SECOND
	NANO	ONE BILLIONTH	NANOSECOND (ns) = ONE BILLIONTH OF A SECOND
	PICO	ONE TRILLIONTH	PICOSECOND (ps) = ONE TRILLIONTH OF A SECOND
MORE THAN 1	KILO	ONE THOUSAND	KILOHERTZ (kHz) = ONE THOUSAND HERTZ (CYCLES PER SECOND)
	MEGA	ONE MILLION	MEGAHERTZ (MHz) = ONE MILLION HERTZ
	GIGA	ONE BILLION	GIGAHERTZ (GHz) = ONE BILLION HERTZ
	TERA	ONE TRILLION	TERAHERTZ (THz) = ONE TRILLION HERTZ

TABLE 1.2 CONVERSIONS FROM PARTS PER... TO PERCENTS

PARTS PER			PERCENT
1 PART PER HUNDRED	= 1×10^{-2}	= $\frac{1}{100}$	1.0%
1 PART PER THOUSAND	= 1×10^{-3}	= $\frac{1}{1,000}$	0.1%
1 PART PER 10 THOUSAND	= 1×10^{-4}	= $\frac{1}{10,000}$	0.01%
1 PART PER 100 THOUSAND	= 1×10^{-5}	= $\frac{1}{100,000}$	0.001%
1 PART PER MILLION	= 1×10^{-6}	= $\frac{1}{1,000,000}$	0.0001%
1 PART PER 10 MILLION	= 1×10^{-7}	= $\frac{1}{10,000,000}$	0.00001%
1 PART PER 100 MILLION	= 1×10^{-8}	= $\frac{1}{100,000,000}$	0.000001%
1 PART PER BILLION	= 1×10^{-9}	= $\frac{1}{1,000,000,000}$	0.0000001%
1 PART PER 10 BILLION	= 1×10^{-10}	= $\frac{1}{10,000,000,000}$	0.00000001%
1 PART PER 100 BILLION	= 1×10^{-11}	= $\frac{1}{100,000,000,000}$	0.000000001%
1 PART PER 1,000 BILLION (1 PART PER TRILLION)	= 1×10^{-12}	= $\frac{1}{1,000,000,000,000}$	0.0000000001%
1 PART PER 10,000 BILLION	= 1×10^{-13}	= $\frac{1}{10,000,000,000,000}$	0.00000000001%

1.7.2 FREQUENCY

The term used almost universally for frequency is the hertz, which means one cycle per second. With the advent of new integrated circuits, it is possible to generate that "cycle" in many different shapes. It can be a sine wave or a square, triangular, or sawtooth wave. So the reader is cautioned as he proceeds through this book to keep in mind that the waveforms being considered may not in fact be sinusoids (sine waves). In this book we will not concern ourselves greatly with the possibility that the frequency of interest can contain zero frequency or DC components. Instead we will assume that the waveform operates, on the average, near zero voltage.

Throughout this book, consideration is given to frequencies of all magnitudes--from the one hertz tick of the clock to many billions of hertz in the microwave region. Table 1.3 gives the prefixes used for frequencies in different ranges and also the means of converting from one kind of unit to another. Thus it is possible to refer to one thousand on the AM radio band as either 1000 kHz or 1 MHz.

Table 1.4 lists the frequencies by bands. Most frequencies of interest are included in this table--the radio frequency band contains the often heard references to high frequency, very high frequency, low frequency, etc.

The reader is cautioned that the difficulty in measuring frequency accurately is not directly related to the frequency range. Precise frequency measurements at audio frequencies are equally as difficult as those in the high frequency radio bands.

Scattered throughout this book are references to the wavelength rather than the frequency being used. Wavelength is especially convenient when calculating antenna

TABLE 1.3 CONVERSIONS TO HERTZ

FREQUENCY	EXAMPLES
1 Hz = 10^0 = 1 CYCLE PER SECOND	2000 Hz = 2 kHz = 0.002 MHz
- 1 kHz = 10^3 = 1,000 Hz	
1 MHz = 10^6 = 1,000,000 Hz	25 MHz = 25,000 kHz = 25 MILLION Hz
1 GHz = 10^9 = 1,000,000,000 Hz	10 GHz = 10,000 MHz = 10 MILLION kHz
1 THz = 10^{12} = 1000 BILLION Hz	

TABLE 1.4 RADIO FREQUENCY BANDS

RF BAND	FREQUENCY RANGE	WAVELENGTH (λ)
4 VLF (VERY LOW FREQUENCY)	3 - 30 kHz	$10^5 - 10^4$ METERS
5 LF (LOW FREQUENCY)	30 - 300 kHz	$10^4 - 10^3$ "
6 MF (MEDIUM FREQUENCY)	300 kHz - 3 MHz	$10^3 - 10^2$ "
7 HF (HIGH FREQUENCY)	3 - 30 MHz	$10^2 - 10$ "
8 VHF (VERY HIGH FREQUENCY)	30 - 300 MHz	$10 - 1$ "
9 UHF (ULTRA HIGH FREQUENCY)	300 MHz - 3 GHz	$1 - 0.1$ "

lengths. In fact, a glance at an antique radio dial shows that the band was actually marked in wavelengths. For example, radio amateurs still refer to their frequency allocations in terms of the 20, 10, or 2 meter bands.

The conversion of wavelengths to frequency can be made for most purposes by using the simple equation

$$\lambda = \frac{300,000,000}{f} \ ,$$

where λ is the wavelength in meters, 300,000,000 meters per second is the speed of light, and f is the frequency in hertz. So we can see that the ten meter band is approximately 30 MHz and 1000 on the broadcast band is 300 meters.

This equation can be converted to feet and inches for ease in cutting antennas to exact length. Precise calculations of wavelength would have to take into account the medium and allow for the reduced velocity below that of light, for instance, inside coaxial cables.

Many users of frequency generating devices tend to take frequency for granted, especially in the case of crystals. A popular feeling is that if the frequency of interest has been generated from a quartz crystal, it cannot be in error by any significant amount. This is simply not true. Age affects the frequency of all quartz oscillators. Although frequency can be measured more precisely than most phenomena, it is still the responsibility of the calibration laboratory technician or general user to keep in mind the tolerances needed; for example, musical notes are usually measured to a tenth of a hertz or better. Power line frequency is controlled to a millihertz. If you dial the NBS time-of-day telephone service, you will hear audio frequency tones that, although generated to an accuracy of parts per one thousand billion, can be sent over the telephone lines only to a few parts in one thousand. So the responsibility is the users to decide what he needs and whether, in fact, the measurement scheme he chooses will satisfy those needs.

Throughout this book, mention is made of the ease with which frequency standards can be calibrated to high precision. Let us assume that you have, in fact, just calibrated your high quality crystal oscillator and set the associated clock right on time. What happens next? Probably nothing happens. The oscillators manufactured today are of excellent quality and, assuming that a suitable battery supply is available to prevent power outages, the clock could keep very accurate time for many weeks. The kicker in this statement is, of course, the word "accurate." If you want to maintain time with an error as small or smaller than a microsecond, your clock could very easily have that amount of error in a few minutes. If you are less concerned with microseconds and are worried about only milliseconds or greater, a month could elapse before such an error would reveal itself.

The point to be made here is that nothing can be taken for granted. If you come into your laboratory on Monday morning hoping that everything stayed put over the weekend, you might be unpleasantly surprised. Digital dividers used to drive electronic clocks do jump occasionally, especially if the power supplies are not designed to avoid some of the glitches that can occur. It makes sense, therefore, to check both time and frequency periodically to insure that the frequency rate is right and the clock is on time.

Many users who depend heavily on their frequency source for calibrations—for example, manufacturing plants—find it convenient to maintain a continuous record of the frequency of their oscillators. This usually takes the form of a chart recording that shows the frequency variations in the oscillator versus a received signal from either an NBS station or one of the many other transmissions which have been stabilized. Among these are the Omega and Loran-C navigation signals.

If you refer to Chapter 11 of this book, which deals with the characteristics of oscillators, you will notice that crystal oscillators and even rubidium oscillators will drift in frequency so, depending on your application, periodic adjustments are required.

1.8 DISTRIBUTING TIME AND FREQUENCY SIGNALS VIA CABLES AND TELEPHONE LINES

Users of time and frequency signals sometimes want to distribute either a standard frequency waveform or a time signal using cables. This is often the case in a laboratory or manufacturing plant where the standards laboratory provides signals for users throughout the plant area. Of course, the solution to any given problem depends on good engineering practices and consideration for the kind of result desired. That is to say, if you start with a cesium oscillator and want to maintain its accuracy throughout a large area, you must use the very best of equipment and cables; and even then, the signal accuracy will be somewhat deteriorated. On the other hand, if all that is needed is a time-of-day signal on the company telephone switchboard, the specifications can be relaxed.

Without knowing the particular conditions under which signals are to be distributed, it would be hard to specify the maximum accuracy obtainable. A number of articles have been published for distribution systems using both coaxial cables and telephone lines. The results that the individual writers reported varied from parts per million to almost one part in 10 billion. In each case the results were in direct ratio to the amount of effort expended. Since the articles were written a number of years ago, it is expected that better results could be achieved today. Other than the good engineering practices mentioned above, there is no simple formula for successful transmission of standard frequency and time signals.

Often the main problem is noise pickup in the cables that causes the signal at the end of the line to be less useful than desired. Another problem often reported is the difficulty in certifying the received accuracy or to establish NBS traceability over such a distribution system. A suggestion would be that careful system management be observed and that periodic evaluation of the system be attempted. One method of evaluating a distribution system would be to route the signal on a continuous loop and look at the signal going in and the signal coming out for a particular cable routing.

For commercial telephone lines (which are usually balanced systems), the highest practical frequency that can be distributed is 1000 Hz. Coaxial cables have been used successfully for frequencies up to 100,000 Hz.

Any attempt to send pulses over a voice-grade telephone line would be unsuccessful due to the limited bandwidth.

At the NBS Boulder Laboratories, 5 MHz signals are transmitted within the building. Commercial distribution amplifiers with fail-safe provisions are used to drive the cables at a nominal 50 ohms impedance. Careful cable management assures good results. Even so, very often noise will appear on the cables and degrade the signals. The noise on some cables due to their routing was so bad they were abandoned.

The NBS telephone time-of-day signal is received at Boulder via a leased commercial telephone line that is about 100 miles in length. The measured delay for this line is approximately 3 to 5 milliseconds. This seems to be the usual order of magnitude of delay reported for commercial telephone circuits. If you plan to use such circuits for distribution of signals, keep in mind that a variety of leased lines of varying quality are available from telephone companies.

For signals that are being transmitted over long distances on telephone lines, the error in the received frequency is usually several hertz or more. This is a consequence of the method used by the telephone companies to combine several signals on a transmission line using frequency division multiplexing techniques. This would be the case, for example, if you dialed the NBS time-of-day service on 303-499-7111. The tone as received could be in error by several parts in 10^3.

For those users wanting to distribute either very accurate frequency or precise time, it is almost fair to say that local cable and wire systems are not economically practical. If your requirements are critical, it is usually cheaper in the long run to simply reestablish time and/or frequency at the destination point. The burden of installing and maintaining a distribution system can be very great for high accuracy systems.

1.9 TIME CODES

Whenever time is available from a digital clock at one location and needed at another, it is often transferred over wires or radio by means of a time code. A time code is a series of pulses which can easily be a simple teletypewriter code, but is usually a more efficient binary code where a set of pulses represents one digit. Let's say a 4 is sent, meaning 4 hours, 4 minutes, or 4 seconds. The location of a particular binary digit in the code tells you its meaning; that is, whether it is an hour, minute, or second. Depending on the application, the code can be sent as a direct current level shift or as modulated pulses on a carrier or perhaps as tones where one frequency of tone represents a binary "1" and an alternate tone represents a binary 0.

What has been described above is a serial code where one digit is sent, then another, then another. Interspersed among the time bits are position locators which help the electronic equipment to recognize what the following bit is going to mean.. It is also possible to send time codes in parallel on many conductor cables. Each wire would then carry its respective bit.

Time codes have evolved through the years and have been strongly influenced by the recommendations of the Inter Range Instrumentation Group, known as IRIG. Figure 1.2 shows a typical IRIG time code format which differs only slightly from the WWV time code shown elsewhere in this book. Figure 1.3 shows the major characteristics of a number of IRIG codes. For those readers who want more information on time codes, many manufacturers of digital timekeeping equipment provide booklets listing all of the popular codes being used and detailed explanations of the individual formats.

If you look in a book listing the available time codes, you will notice that they vary in frame length and rate. The NBS time codes were chosen as a compromise on speed and length that matched the other radio station characteristics. The codes on the NBS stations are transmitted very slowly--so slowly that they can be recorded directly on chart

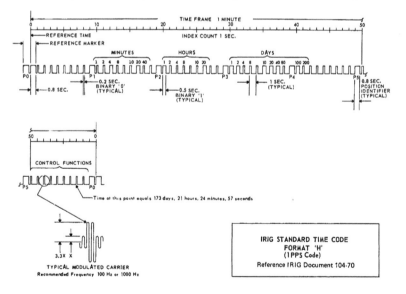

FIGURE 1.2. IRIG TIME CODE, FORMAT H.

13

FORMAT A

1000-PPS, 34-BIT BCD TIME-OF-YEAR/17-BIT BINARY TIME-OF-DAY CODE

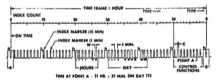

TIME AT POINT A   21:18:42 · .8 · .07 · .005   21 HR. 18 MIN. 42.875 SEC. ON DAY 173

FORMAT B

100-PPS, 30-BIT BCD TIME-OF-YEAR/17-BIT BINARY TIME-OF-DAY CODE

TIME AT POINT A · 21:18:42 · .7 · .05 · 21 HRS. 18 MIN. 42.75 SEC. ON DAY 173

FORMAT D

1-PPM, 16-BIT BCD TIME-OF-YEAR CODE

TIME AT POINT A · 21 HR. · 57 MIN. ON DAY 173

FIGURE 1.3. IRIG TIME CODE FORMATS.

recorders or read by ear if necessary. CHU in Canada and other time and frequency stations also have time codes as part of their broadcast format.

The time codes provided on the NBS radio stations WWV, WWVH, and WWVB can be used for clock setting. But because of fading and noise on the radio path, these pulses cannot usually be used directly for operating a time display. Instead, the information is received and decoded and used to reset a clock that is already operating and keeping time. However, the noise and fading often cause errors in the decoding so any clock system that intends to use radio time codes must have provisions to flywheel with a crystal oscillator and maintain the required accuracy between the successful decodes.

Does this mean that the time codes as transmitted and received are of little prac-

tical value? No, on the contrary. There are so few sources of accurate, distributed time that these time codes are of immense value to people who must keep time at remote locations. The description of how to use the time codes would suggest that they are complicated and cumbersome. This may have been true a few years ago. But with the advent of modern integrated circuit technology, both the cost and complexity have been reduced to manageable proportions. The real point to be noticed is that the code cannot be used directly but must be used to reset a running clock with suitable data rejection criteria.

Information obtained by the NBS from its mail and direct contact with users suggests that time codes will be used more and more. Several factors have strongly influenced this increased usage. One of these is the availability of low-cost digital clocks. The other

14

factor, as previously mentioned, seems to be the cost reductions in integrated circuits. These circuits are so inexpensive and dependable that they have made possible new applications for time codes that were impractical with vacuum tubes.

The accuracy of time codes as received depends on a number of factors. First of all, you have to account for the propagation path delay. A user who is 1000 miles from the transmitter experiences a delay of about 5 microseconds per mile. This works out to be 5 milliseconds time error. To this we must add the delay through the receiver. The signal does not instantly go from the receiving antenna to the loudspeaker or the lighted digit. A typical receiver delay might be one-half millisecond. So a user can experience a total delay as large as 8 to 10 milliseconds, depending on his location. This amount of delay is insignificant for most users. But for those who do require accurate synchronization or time of day, there is a method for removing the path delay. Manufacturers of time code receivers very often include a simple switching arrangement to dial out the path delay. When operating time code generators, the usual recommendations for battery backup should be followed to avoid errors in the generated code.

THE UTILITY OF TIME CODES. In addition to using codes for setting remote clocks from, say, a master clock, time codes are used on magnetic tape systems to search for information. The time code is recorded on a separate tape track. Later, the operator can locate and identify information by the time code reading associated with that particular spot on the tape. Many manufacturers provide time code generation equipment expressly for the purpose of tape search.

Another application of time codes is for dating events. The GOES satellite time code (chapter 10) is used to add time and a date to environmental data collected by the satellite.

There are many other uses. The electric power industry automatically receives the time code from WWVB, decodes the time of day, and uses this information to steer the electric power network. Many of the telephone companies use a decoded signal to drive their machines which actually answer the phone when you dial the time of day. Even bank signs that give time and temperature are often driven from the received time code. Why is this? The main reason a time code is used is to avoid errors. It also reduces the amount of work that must be done to keep the particular clock accurate.

15

CHAPTER 2.   THE EVOLUTION OF TIMEKEEPING

2.1 TIME SCALES

A time scale is simply a system of as-
signing dates to events or counting pendulum
swings. The apparent motion of the sun in the
sky is one type of time scale, called astronomical time. Today, we also have atomic
time, where an atomic oscillator is the "pendulum". To see how the various time scales
came about, let's take a brief look at the
evolution of time.

2.1.1 SOLAR TIME

To review what has happened
to timekeeping and the name giv-
en to the resulting time of day,
consider a time system that uses
the sun and the sundial. The sun
is the flywheel and has a period
of 24 hours. The sundial can
indicate the fractions of cycles
(time of day). As complete days
elapse, calendars can be used to
count the days and divide them
into weeks and years. But our
newly formed clock is not uniform in its timekeeping because
the earth's orbit around the sun
is not circular. It slows down
and speeds up depending on its
distance from the sun. The early astronomers and mathematicians understood these laws of
motion and were able to correct
the "apparent solar time" to obtain a more uniform time called
"mean solar time". This correction is called the Equation of
Time and is often found engraved
on sundials. Universal Time
(UTO) is equal to mean solar time
if you make the correction at the
Greenwich meridian in England.
We now have UTO, the first in a series
of designations for time that has evolved
through the past years.

If you use a star that is farther away
than our own sun, the fact that the earth is
not in perfect circular orbit becomes unim-
portant. This is called "sidereal time" and
is similar to mean solar time since both are
based on the earth's spinning on its axis.
The rate is different by one day per year
since the earth circles the sun in a year.

As better clocks were developed, astronomers began to notice a discrepancy in Universal Time measured at different locations. The

discrepancy was eventually identified as being
caused by a wobble in the earth's axis. The
amount of wobble is about 15 meters. By careful measurements made at various observatories
throughout the world, this wobble was corrected for and a new time designation called UT1
was born. In our search for uniformity, we
have now taken care of the non-circular orbit
(UTO) and the axis wobble of the earth (UT1).

As science and technology improved pendulum and quartz crystal clocks, it was dis-

covered that UT1 had periodic fluctuations
whose origin was unknown. Due to the availability of excellent clocks, these fluctuations
could be and were removed, resulting in an
even more uniform UT2.

To review, UT1 is the _true_ navigator's
scale related to the earth's angular position.
UT2 is a smooth time and does not reveal the
real variations in the earth's position. When
the world's timekeepers went to UT2, they
bypassed the navigators' real needs. A little
later we shall describe the present-day system
which in effect remedies the dilemma.

Up until now we have been talking about the Universal Time family, and figure 2.1 shows the relationship between the different universal times. Let us now examine the other members of the time family. The first of these is "ephemeris time". An ephemeris is simply a table that predicts the positions of the sun, moon, and planets. It was soon noticed that these predicted positions on the table did not agree with the observed positions. An analysis of the difficulty showed that in fact the rotational rate of the earth was not a constant, and this was later confirmed with crystal and atomic clocks. In response to this problem, the astronomers created what is called "ephemeris time". It is determined by the orbit of the earth about the sun, not by the rotation of the earth on its axis.

UNIVERSAL TIME FAMILY

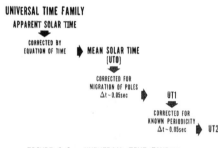

FIGURE 2.1.   UNIVERSAL TIME FAMILY
RELATIONSHIPS.

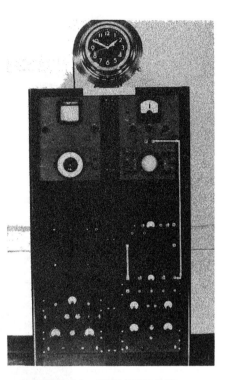

FIGURE 2.2.   FIRST ATOMIC CLOCK.

2.1.2 ATOMIC TIME

Another kind of time that can be generated and used is atomic time. Whereas the Universal Time scale is obtained by counting cycles of the earth's rotation from some agreed-upon starting point, an atomic time scale can be obtained by counting cycles of a signal locked to some atomic resonance.

In the late 1940's, the National Bureau of Standards announced the first atomic clock. Shortly thereafter several national laboratories, including the International Time Bureau (BIH), started atomic time scales.

In 1971 the General Conference of Weights and Measures officially recognized the Atomic Time Scale and endorsed the BIH time scale as the International Atomic Time Scale, TAI. The role of the BIH in international coordination is discussed later in this chapter. Since the first of January 1972, the Atomic Time Scale has been distributed by most countries in the world.

Today, the time of day can be gotten to a very high order of accuracy because, unlike a coarse sundial, an atomic clock measures small fractions of a second. Furthermore, it gives us essentially constant units of time--it is a uniform timekeeper. Uniformity is important in technology, synchronization of power generators, and in the general problem of trying to make two things happen or start at the same time.

Let's review for a moment the several time scales we have discussed. First, the universal time family is dependent on the earth's spin on its axis. Second, ephemeris time depends on the orbital motion of the earth about the sun. And finally, atomic time, which is very uniform and precise, depends on a fundamental property of atoms.

Figure 2.3 illustrates the differences in these several kinds of time. Because of the slow orbital motion of the earth, about one

18

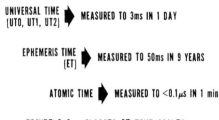

UNIVERSAL TIME ➤ MEASURED TO 3ms IN 1 DAY
(UT0, UT1, UT2)

EPHEMERIS TIME ➤ MEASURED TO 50ms IN 9 YEARS
(ET)

ATOMIC TIME ➤ MEASURED TO <$0.1\mu s$ IN 1 min

FIGURE 2.3.   CLASSES OF TIME SCALES
AND ACCURACIES.

cycle per year, measurement uncertainties limit the accuracy of Ephemeris Time to about 0.05 second in a 9-year period. Universal Time can be determined to a few thousandths of a second or several milliseconds in one day. Atomic Time is accurate to a few billionths of a second in a minute or less. From these numbers and the figure, it is easy to see why scientists have been leaning toward a time scale derived from an atomic clock.

A.   Coordinated Universal Time

Prior to 1972, most standard frequency radio broadcasts were based on a time scale called Coordinated Universal Time (UTC). The rate of a UTC clock was controlled by atomic oscillators so it would be as uniform as possible. These atomic oscillators operated at the same frequency for a whole year, but were changed in rate at the beginning of a new year in an attempt to match the forthcoming earth rotational rate, UT2. This annual change was set by the BIH. For instance, between 1965 and 1966, the relative frequency, F, of standard frequency radio broadcasts was set equal to -150×10^{-10}. This meant that all of the WWV frequencies, for example, were lower than their nominal values. (See Chapter 3 for a full explanation of nominal values.)

However, the earth's rotational rate could not be accurately predicted, and so UTC would slowly get out of step with earth time. This was a problem for navigators who needed solar time--they had to apply a correction to UTC, but it was difficult to determine the amount of correction.

Experience rapidly showed that it would be an advantage to simplify the UTC system to avoid changing the rate each year. A decision was made to broadcast the nominal values of the frequency standards and to do away with annual changes in rate. Thus, a way was developed to keep UTC in closer agreement with solar time.

B.   The New UTC System

The new UTC system came about as a solution to the problem of changing the frequency rate each year or so. It was adopted in Geneva in 1971 and became effective in 1972. Under the new system, the frequency driving all clocks was left at the atomic rate with zero offset. But by using a zero offset, the clocks would gradually get out of step with the day. The resulting situation was not unique because the year has never been an exact multiple of the day, and so we have leap years. Leap years keep our calendar in step with the seasons.

The same scheme was adopted to keep clocks in step with the sun, and the "leap second" was born. To make adjustments in the clock, a particular minute would contain either 61 or 59 seconds instead of the conventional 60 seconds. You could, therefore, either have a positive or a negative leap second. It was expected and proved true that this would normally occur once a year.

The new UTC plan works like this. By international agreement, UTC is maintained within 9/10 of a second of the navigator's time scale UT1. The introduction of leap seconds will permit a good clock to keep approximate step with the sun. Since the rotation of the earth is not uniform we cannot exactly predict when leap seconds will be added or deleted, but they usually occur on June 30 or December 31. Positive leap seconds were added on June 30, 1972, and December 31, 1972 through 1978.

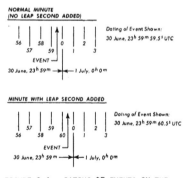

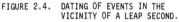

FIGURE 2.4.   DATING OF EVENTS IN THE
VICINITY OF A LEAP SECOND.

What does this mean to the user of a time and frequency broadcast? It simply means that the time he gets will never differ from UT1 by more than 9/10 of a second. Most users, such as radio and television stations and telephone

19

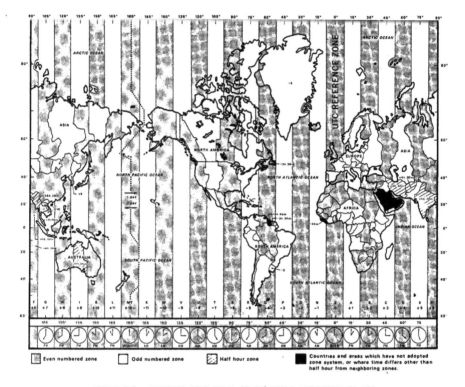

FIGURE 2.5. STANDARD TIME ZONES OF THE WORLD REFERENCED TO UTC.

time-of-day services, use UTC so they don't care how much it differs from UT1. Even most boaters/navigators don't need to know UT1 to better than 9/10 of a second, so the new UTC also meets their needs.

However, there are a small number of users who do need UT1 time to better than 9/10 of a second. To satisfy this need, most standard time and frequency radio stations include a correction in their broadcasts which can be applied to UTC to obtain UT1.

On WWV, for instance, the corrections, in units of 0.1 second, are encoded into the broadcasts by using double ticks or pulses after the start of each minute. The amount of correction is determined by counting the number of successive double ticks heard each minute, and the sign of the correction is

given by the location of the double ticks within the minute.

We should keep in mind that the advantage we gain by using the new UTC system prevents us from simply subtracting the dates of two events to get the time difference between them. We must, in fact, take into account any leap seconds that were added or deleted. You should be especially cautious if the time interval of interest extends backward into a period where the world time system was not operating on the new UTC time scale (prior to 1972).

2.1.3 TIME ZONES

All standard time and frequency stations broadcast Coordinated Universal Time, which is

20

referenced to the Greenwich meridian. However, many users want to display local time. If the time is being decoded from a time code (as opposed to a voice time-of-day announcement), the problem can be solved by using integrated circuit logic. In fact, special chips have been manufactured that allow the clock to display time from any of the world time zones, even though it is receiving and decoding UTC. Figure 2.5 shows a map of the time zones currently in use.

Contrary to popular opinion, the National Bureau of Standards is not involved with determining time zones in the United States. This responsibility has been given to the Department of Transportation because the need for designating time zones came about from the use of railroads for interstate commerce. Information about the time zones in the U. S. can be obtained from the Department of Transportation, Washington, DC 20590.

To review then, we have the unfortunate situation where the scientist wants a uniform time scale (UTC) and the navigators require a clock tied to the earth's position (UT1) which is non-uniform. Therefore, corrections must be made to convert UTC to UT1.

With this discussion, the reader can more easily understand why the new UTC system was developed. It is a compromise between a purely uniform time scale for scientific application and the non-uniform measurement of the earth's position for navigation and astronomy. As mentioned before, we keep the calendar in step with the seasons by using the leap years and our clocks in step with the sun (day and night) by using leap seconds.

2.2 USES OF TIME SCALES

2.2.1 SYSTEMS SYNCHRONIZATION

There are many applications for timing where it does not matter if the clock in use is on time or not, nor does it matter if the seconds are uniform. A child's race is a good example. If someone shouts "Go", the starting time is synchronized and it doesn't matter if it occurred on the second. In other words, it is not essential that clocks be on the right time--it is only important that they all read the same time--that is, they have to be synchronized.

It is only a matter of convenience and convention to encourage the use of the correct time. The convenience arises because of the ease with which lost time can be recovered. If a locomotive engineer can get the time to reset his watch by simply calling his local telephone company or asking another trainman, he is able to maintain his schedule with ease.

In summary, there is an advantage and benefit to keeping everyone synchronized and on the correct time. In a way it is much like the argument in favor of standards themselves. If you stop to help a motorist by the side of the road, your confidence that your tools will fit his car is reassuring. In the same way, by using the same time of day, annoyances caused by errors are reduced.

Before we leave the subject of synchronization, we should mention that there are applications in astronomy, communications, and navigation where synchronizations to an

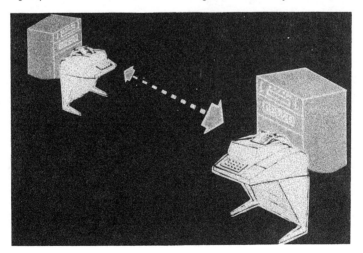

accuracy of a millionth of a second or better are required. Although the average person is not terribly concerned if he arrives at school, church, or a picnic on time to within a few minutes or so, an airliner navigating with a clock in error by one millionth of a second (microsecond) would have a position error of a quarter of a mile.

This is also true internationally with respect to high-speed digital communications where several million characters can be sent and received each second. Here the "time" being used may be a few microseconds, so the sending and receiving equipment must be synchronized to this accuracy. If this synchronization can be obtained from a local navigation or time and frequency station, it enables rapid, error-free data transmission.

2.2.2 NAVIGATION AND ASTRONOMY

Although most users like the uniformity of atomic time, there is one application that needs the non-uniformity of solar time. This is in the area of celestial navigation where the motion of the earth is a way of finding out where you are on the earth.

Navigators who find their positions from the stars are among the greatest users of standard time broadcasts. Since the earth makes one revolution in 24 hours, a person can find his position (in longitude) if he knows his local time and the time difference between the Greenwich meridian and himself.

As an example, a person trying to find his position uses an instrument like a sextant to measure the local solar time wherever he might be on the earth. His problem then is trying to find out what the time is at the Greenwich meridian. This is the same problem that generated all the interest several hundred years ago in developing a good chronometer. These clocks were used for many years until radio came on the scene. Even today, most vessels do not leave home without a good chronometer.

Because of this vital need for time of day on the high seas, many countries operate high frequency radio stations to serve navigators. The United States operates WWV in Fort Collins, Colorado, and WWVH in Hawaii. Similar stations are operated by the Canadians and the Japanese and many countries in Europe and Asia. The U. S. Navy also broadcasts time signals from a number of radio stations.

Most large countries also maintain observatories to measure and record astronomical time. This is done by using telescopes, and the official United States Observatory is operated by the Navy in Washington, D. C. Astronomical time is somewhat difficult to measure accurately. Errors of a few milliseconds are realized only after a whole evening's sightings have been taken and averaged.

THE HARRISON CHRONOMETER.

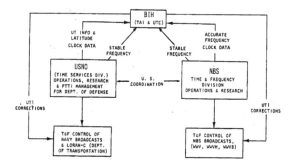

FIGURE 2.6. NBS, USNO, AND BIH INTERACTIONS.

2.3 INTERNATIONAL COORDINATION OF TIME AND FREQUENCY ACTIVITIES

2.3.1 THE ROLE OF THE INTERNATIONAL TIME BUREAU (BIH)

Unlike standards such as length or mass, time of day can be received from many sources. Both Canada and the U. S. have primary frequency standards that act as flywheels for accurate timekeeping. In addition, the U. S. Naval Observatory makes astronomical observations for determining UT1 and keeps accurate clocks running for use by the Department of Defense.

Other laboratories throughout the world also maintain primary frequency standards for their countries: The PTB in Germany, the National Physical Laboratory in England, the Radio Research Labs in Japan, and the National Electrotechnical Institute in Italy, just to name a few.

Time and frequency information from all of these laboratories is collected by the International Time Bureau (BIH) in Paris, France, which is the agency responsible for international coordination. There, the information is evaluated and the actual time corrections for each contributing clock are determined. By international agreement, all UTC time scales must agree with the UTC time scale operated by the BIH to within ± 1 millisecond. The result is a uniform world time system that differs from country to country by only a few millionths of a second. So, whether you get your time from CHU in Canada or the PTB in Germany, it will differ from U. S. time by no more than a few microseconds.

2.3.2 THE ROLE OF THE NATIONAL BUREAU OF STANDARDS (NBS)

The National Bureau of Standards is in the Department of Commerce of the United States and has been authorized by Congress to undertake the following functions: "The custody, maintenance, and development of the national standards of measurements and the provisions of means and methods for making measurements consistent with those standards, including the comparison of standards used in scientific investigations, engineering, manufacturing, commerce, and educational institutions, with the standards adopted or recognized by the Government."

The Time and Frequency Division, located in Boulder, Colorado, is that part of the National Bureau of Standards that carries out the above functions related to time and frequency. There are several groups within the Division, each having different responsibilities. The Frequency and Time Standards Group operates the NBS standard of frequency and time interval, as well as several time scales based upon this standard. In addition to maintaining the standard, research efforts are carried out to improve it. This group also offers a direct service for calibration of oscillators and clocks.

The Time and Frequency Services Group is responsible for distributing the standards and for finding new and improved methods of dissemination. At this writing, the dissemination services consist of high frequency radio stations WWV and WWVH, low frequency radio station WWVB, a telephone time-of-day service, time and frequency calibration services using network television, and a satellite time code. This Group also provides services in the form

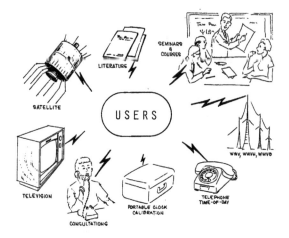

of literature and courses on how to use these services and provides calibration and monitoring data on a regular basis to users of NBS and other time and frequency services.

Research and development into new methods of dissemination are being done in the Time and Frequency Coordination Group. This Group is also involved in national and international time and frequency coordination activities.

A fourth group concerns itself with frequency synthesis techniques and laser stabilization methods.

In addition, services are available in the form of consultations, either by letter, telephone, or visits to the Boulder Laboratories. Users of the time and frequency services are encouraged to call upon us for advice or assistance.

To summarize, the Bureau of Standards maintains services based upon the primary frequency standards. These services are constantly being improved and are designed to meet the needs of users at all levels of accuracy for frequency and time calibration. They range from the telephone time-of-day service to the ultra-stable low frequency radio service of WWVB which is used for precise frequency calibration.

2.3.3 THE ROLE OF THE U.S. NAVAL OBSERVATORY (USNO) AND THE DOD PTTI PROGRAM

The U. S. Naval Observatory (USNO) is in the Department of Defense (DOD). It's mission with reference to time is:

"Make such observations of celestial bodies, natural and artificial, derive and publish such data as will afford to United States Naval vessels and aircraft as well as to all availing themselves thereof, means for safe navigation, including the provision of accurate time." The practical application of this mission is for the USNO to serve as the operational reference for the official United States time of day (as opposed to responsibilities for the unit of time and frequency in NBS).

The Time Service Division of the USNO pursues several activities: It operates Photographic Zenith Tubes (telescopes) to determine Universal Time. It operates a large bank of atomic clocks (more than 20) as the basis of an extremely reliable and uniform atomic time scale. It transports portable atomic clocks worldwide for time synchronization to "assure worldwide continuity of precision". It promptly publishes and distributes Time Service Announcements regarding measurements of, and information on, many electronic systems useful for time dissemination. It operates under contract with National Science Foundation (NSF), a connected radio interferometer at the Green Bank National Radio Astronomical Observatory for the determination of UT with radio techniques.

The USNO contributes measurements, along with about 60 other observatories and standards laboratories throughout the world, to the Bureau International de l'Heure (BIH) in Paris, France, which combines the results into final BIH values of UT1. The USNO is the only organization in the United States that determines UT1 operationally. It distributes BIH Circular D for the BIH in the United States.

24

Within the DOD, the USNO has been given management responsibilities to "accomplish overall PTTI objectives at minimum cost, DOD Directive 5160.51 of 31 August 1971.

The DOD provides frequency and time signals throughout the world. These include many different radio frequencies, satellite signals, and portable clocks. These can provide very high resolution time synchronization, one-tenth microsecond or better at the present time (1979) with 10 nanoseconds possible in the near future. Specific examples include the Transit system with worldwide automatic timing capability to 50 μs (single pass, 1 sigma; averaging over a number of passes can give about 5 μs, 1 sigma).

The Global Positioning System is currently under development and will provide a worldwide timing capability of about 10 ns. Loran-C is operated by the U. S. Coast Guard. It acts in a secondary mission for the dissemination of time for the USNO. It provides time within ground wave coverage to about 1 μs with a precision of about 0.1 μs. Loran-C is the present operational system for the computation and dissemination of the internationally coordinated clock time references TAI and UTC.

The USNO accomplishes its global responsibilities with a system of cooperating "Precise Time Reference Stations (PTRS)". Any of these PTRS clocks can be used to gain access to the USNO Master clock by correcting a reading against the PTRS clock with the published difference USNOMC-PTRS. The currently designated PTRS's and the respective values are given in Time Service Announcement Series 16. Time Service Announcement Series 4 publishes Loran-C, Omega, VLF TV phase (time) values and Time Service Announcement Series 7 has time scale information (atomic and astronomical, including polar coordinates). These bulletins are available on request from:

Superintendent
U. S. Naval Observatory
Attention: Time Service Division
34th & Massachusetts Avenue, NW
Washington, D. C. 20390
(Telephone: 202 254-4546, AV 294-4546)
(TWX 710-822-1970)

The three military services of the DOD have established focal points for matters of Precise Time and Time Interval (frequency). These focal points coordinate PTTI efforts and provide pertinent information to their organizations and contractors. They also arrange portable clock synchronizations for their organizations. The focal points are:

1. Aerospace Guidance and Metrology Center, Newark Air Force Station, Ohio (for the U. S. Air Force).

2. Navy Electronic Systems Command, Code 510, Washington, DC 20360 (for the Navy).

3. U. S. Army Material Development & Readiness Command, DRCQA-PC, Alexandria, VA 20315 (for the Army).

Users of time and frequency signals, equipment, and publications need to have a language for keeping records, reporting performance, and reading the published literature. It should be easily and correctly understood. Thus, a convention is needed for the algebra dealing with frequency sources and clocks so that the meaning of data are understandable and not confusing.

You might ask, "Hasn't this problem come up before?" or "How has it been resolved?" Yes, it has come up before and it has been resolved. The CCIR (International Radio Consultative Committee) has defined the problem and published a recommended, standard notation. In this section, we will explain the notation with examples. Notice that this convention regarding the algebra and especially the algebraic signs has been followed throughout this book.

In many cases of time and frequency measurement, the user knows that his clock is fast or that his frequency source is low with respect to NBS, the USNO, or another oscillator. Intuition is a good guide and the careful worker will be able to confirm his suspicions by reference to his data and the published information he needs. But in some cases, intuition is not enough and the adoption of a standard notation will eliminate difficulties. Annotation of data sheets or charts should follow the CCIR recommendations so that a person looking up something that happened several months or years ago will not be confused.

The CCIR convention and notation is listed here and can also be found in the publications of that group. The following material was excerpted from Recommendation No. 459, 1970. The material has been rearranged for clarity.

1. Algebraic quantities should be given to avoid confusion in the sign of a difference in the time between clocks or the frequency of sources.

2. It is a good idea to give the date and place of a clock reading or frequency measurement.

3. The difference of the readings (a and b) of clocks A and B is, by convention:

$$A - B = a - b.$$

An example using the above equation would be:

$$A_{1pps} - B_{1pps} = 5h\ 6m\ 2s - 5h\ 6m\ 1s = +1s.$$

This time difference expression is ideally suited to the kind of readings obtained with time interval counters. This counter is explained in Chapter 4 and produces a readout by having one pulse open the counter gate or START the count and another close the gate, STOPPING the count. In the above equation, clock A is the start pulse and clock B is the stop pulse.

Can we ever get a negative algebraic sign on the right-hand side of the CCIR equation? Based directly on the counter measurements, we cannot because, if the pulse from clock B occurs 50 microseconds before the pulse from clock A, the counter will simply wait for the next pulse and read 999,950 microseconds. It does not display negative numbers at all!

Remember, these clocks are ticking once per second and our counter looks only at this seconds tick--it cannot see the minutes or hours. We say then that our clocks have outputs that are ambiguous to one second. This means that the counter reading of 999,950 microseconds is equivalent to -50. It means the same thing. The problem of ambiguity is illustrated below.

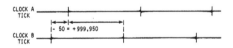

Another example, showing the correct notation that should be used, follows. This could be the result obtained by carrying an NBS portable clock to the U. S. Naval Observatory and measuring it against the USNO clock. The results of this comparison are properly written:

NBS PORTABLE CLOCK - UTC(USNO) =

+12.3 MICROSECONDS

(7 JULY 68, 14h 35m UTC).

There can be no confusion about the meaning of this measurement providing the whole expression is recorded. Do not simply make a claim that the NBS portable clock leads, lags, is fast or slow, or ahead or behind of the USNO clock. The algebraic expression should be used!

It is also important to follow the notation for frequency differences. Terms like high, low, and offset should be avoided. Since the algebra speaks for itself, it should be used with a minimum of additional terms.

3.1 EXPRESSING FREQUENCY DIFFERENCE (IN HERTZ, kHz, MHz, ETC.)

The actual frequency of a source is indicated by the lowercase letter, f. (Refer to the glossary at the end of this book for a further explanation and definition.) A subscript can be used to identify the particular frequency source; e.g., f_c for an oscillator designated "c." Multiple letter subscripts are sometimes used as in the case of NBS, USNO, ABC, NBC, and CBS. With this notation, we can write:

$$f_a - f_b = \Delta f,$$

where Δf is the frequency difference in hertz, kHz, MHz, etc.

The significance of this seemingly simple expression must be stressed. This is the <u>actual</u> frequency coming out of the source (e.g., oscillator) and not the nominal frequency, f_n, that is written on the manufacturer's nameplate or in the specifications. However, it is useful to note the nominal frequency, f_n, of the source when writing down this measurement result.

3.2 RELATIVE (FRACTIONAL) FREQUENCY, F (DIMENSIONLESS)

The term "fractional," though in common use throughout the English-speaking world, is being discouraged by the CCIR because of difficulties in translating that term into certain other languages. The CCIR prefers the use of "relative" or "normalized" rather than

"fractional." We have reflected this throughout the text.

The relative frequency departure of frequency source "c," for instance, from its nominal value f_{nc} is defined as:

$$F_c = \frac{F_c}{f_{nc}} - 1 \text{ (which is dimensionless),}$$

where F is the designation for relative frequency. This expression can be rearranged:

$$F_c = \frac{(f_c - f_{nc})}{f_{nc}} \text{ (dimensionless)}$$

It now has the form $\Delta f/f$ which is often used in frequency measurements and computations. Be very careful to note the exact order of the terms in the numerator!

3.3 RELATIVE (FRACTIONAL) FREQUENCY DIFFERENCE, S (DIMENSIONLESS)

We can apply the formula in the above equation to two separate frequency sources and then take a difference in that quantity. It will be the relative frequency difference between two sources. For example, consider sources NBS and USNO.

$$S = F_{NBS} - F_{USNO},$$

where S is the relative frequency difference.

Let's assume that an NBS oscillator was measured against a USNO oscillator. The results could be written as:

$$F_{NBS} - F_{USNO} = +5 \times 10^{-13} \text{ (7 July 1968,}$$

$$14 \text{ h } 35 \text{ m to } 20 \text{ h } 30 \text{ m UTC; USNO).}$$

This, then, is a complete, unambiguous statement about the frequencies of two sources. There is a general temptation to use the resulting algebraic signs from such calibrations and make statements about such and such a frequency source being low or high relative to another. The examples using this notation shown below illustrate the difficulty in making such statements.

3.4 EXAMPLE OF ALGEBRA AND SIGN CONVENTIONS IN TIME AND FREQUENCY MEASUREMENTS

3.4.1 TELEVISION FREQUENCY TRANSFER MEASUREMENTS (covered in Chapter 7)

The NBS TIME AND FREQUENCY BULLETIN reports the measured values of the three major U. S. networks. On page 149 of this book, a portion of that bulletin shows typical data obtained from the TV measurements.. Let's look more closely at this data.

The heading for the data is labeled "RELATIVE FREQUENCY (in parts in 10^{11})." Following the CCIR convention, then, these data represent F_{ABC}, F_{NBC}, and F_{CBS}.

Expressing the first NBC reading in equation form, using the $\Delta f/f$ arrangement:

$$F_{NBC} = \frac{f_{NBC} - f_{NBC(nominal)}}{f_{NBC(nominal)}} = -3019.4 \times 10^{-11}.$$

What does this tell us? If we assume we are dealing with a frequency source at 5 MHz, we can substitute that number for the nominal frequency and solve for the actual frequency of the signal being received from NBC via the TV color subcarrier:

$$F_{NBC} = \frac{f_{NBC} - f_{NBC(nominal)}}{5 \times 10^6} = -3019.4 \times 10^{-11}.$$

Using the equality on the right:

$$f_{NBC} - (5 \times 10^6) = -3019.4 \times 10^{-11} \times 5 \times 10^6,$$

$$f_{NBC} - (5 \times 10^6) = -15097 \times 10^{-5},$$

$$f_{NBC} = (5 \times 10^6 - 0.15097) \text{ Hz}.$$

The solution to the above equation yields:

$$4,999,999.84903 \text{ Hz}$$

as the frequency of the NBC oscillator.

This computation reveals that the NBC frequency source is lower in frequency than its nominal value of 5 MHz.

Let's apply the above equation to the NBS frequency standard itself. By definition, since NBS has the standard of frequency, there can be no average difference between its nominal value and its actual value. So:

$$F_{NBS} = 0.$$

If we were to compute the relative difference in frequency between NBS and NBC, the solution would indicate:

$$S_{NBS-NBC} = F_{NBS} - F_{NBC} = 0 - (-3019.4 \times 10^{-11}),$$

$$S = +3019.4 \times 10^{-11}.$$

3.4.2 TELEVISION LINE-10 TIME TRANSFER MEASUREMENTS (covered in Chapter 8)

Let's take an example using data obtained from time measurements. Here, the CCIR convention, while clear enough, needs to be applied with more care than usual. This is because people use intuition more readily when dealing with time measurements. They feel at home with fast, slow, early, and late, and are prone to make mistakes.

Using the example on page 150 of this book for the TV Line-10 measurements, the user is expected to find the relative frequency for his own oscillator. Notice that this problem involves the use of time measurements to compute frequency! Figure 8.7, column one, gives time difference data of:

29

UTC(NBS) - LINE-10(NBC)

= 02216.54 microseconds,

which follows the CCIR convention and indicates that a counter was started with the pulse from the NBS clock and stopped with the next Line-10 pulse. The data in column two represent the same measurement made at a user's location, giving:

User Clock - Line-10(NBC)

= 05339.9 microseconds.

From the above, column three is obtained, by subtraction, giving:

UTC(NBS) - User Clock

= -3123.4 microseconds.

Since the user wants to know where his clock is with respect to NBS, by convention the NBS clock is listed first in the equation. The minus sign (-) indicates that the user's clock pulse occurred before the NBS clock pulse. This can be illustrated graphically (as it would appear on an oscilloscope, for example):

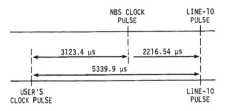

It also illustrates how Line-10 (or any common time pulse) transfers information from NBS to the user. To get the exact time error between the two clocks, path and equipment delays must be taken into account. The final result could easily change the sign from (-) to (+) since path delays could contribute many thousands of microseconds.

But what about the frequency of the oscillator used to drive the user's clock? Is it high or low in frequency with respect to NBS?

To use our derived relationship between frequency and time measurements, we need the change in time over the interval being considered. As shown in figure 8.7, the clock being measured moved -42.5 microseconds in four days, or an average of -10.6 microseconds per day. (This information will give us the needed $\Delta t/T$.) Now,

$$\frac{\Delta f}{f} \doteq - \frac{\Delta t}{T},$$

$$\frac{\Delta f}{f} = -(- \frac{10.6}{86,400 \times 10^6}) = 1.23 \times 10^{-10}.$$

What does this mean? According to the CCIR convention this is

$$F_{user's\ oscillator}$$

or, stated correctly, the oscillator being measured using Line-10 TV measurements has a fractional or relative frequency of +1.23 $\times$ 10^{-10}. The plus sign (+) on F indicates that the oscillator output frequency is higher than its nominal value.

The actual frequency of the oscillator can be calculated from the CCIR definition using:

$$F = \frac{f - f_n}{f_n},$$

and if we assume that the oscillator has a nominal frequency of 1 MHz, its actual frequency can be computed. In this case, it is 1,000,000.000123 Hz. If the nominal frequency of the oscillator being tested were 2.5 MHz, its actual frequency would be

2,500,000.0003075 Hz.

The next page gives the derivation of our equations and an example.

3.5 USING TIME TO GET FREQUENCY

In our search for a method of relating time measurements to frequency measurements, we would want to know how far a particular clock has drifted in a day. Notice that this is a result that can be written, as micro-seconds per day. It tells us the form of the answer we are seeking. We want to know the change in time (microseconds) during a spec-ified period (day). If the above were ex-pressed mathematically, it would be in the form,

$$\frac{\Delta t}{T},$$

To this we want to relate our frequency mea-surements. Although we could, we typically do not measure the frequency error of our oscil-lator in hertz. It is usually a percentage or a ratio. This method of rating oscillators is very common and results in having the numbers come out as numerics, that is, without the dimension of hertz. Algebraically, the fre-quency measurements have the form:

$$\frac{\Delta f}{f},$$

Our task is to relate our $\Delta f/f$ frequency mea-surements to our $\Delta t/T$ time measurements.

This can be done with calculus and is shown below. If the reader is not familiar with calculus, he may want to skip to the result, that states:

$$\frac{\Delta f}{f} = -\frac{\Delta t}{T}.$$

Before we get to the derivation of this equation by using calculus, let's look at a simple example that is not too rigorous, but might offer encouragement that the math works. We have an oscillator with a nominal frequency of 1 MHz, but it is actually running at 999,999 hertz. Let's find

$$\frac{\Delta f}{f} \text{ and } \frac{\Delta t}{T}$$

and see if they agree with the above equation. Clearly, Δf is -1 Hz, both by inspection and by the CCIR definition of $\Delta f = f - f_n$. Since $f = 1$ MHz, then

$$\frac{\Delta f}{f} = \frac{-1}{1 \times 10^6} = -10^{-6}.$$

If we look at this 1 MHz oscillator in-stalled in a clock, we note that it will not "tick" once per second, but at a slightly dif-ferent rate. Let's apply the CCIR formula for clocks to this one and let's further assume it is being compared to the NBS clock. If we start the clocks together, their difference will be zero, but a second later:

NBS CLOCK - 1 MHz CLOCK = 1 MICROSECOND.

Ten seconds later, we will have a result that is 10 microseconds and so on. We now compute $\Delta t/T$. It is one microsecond per second or $10^{-6}/1$. This agrees with the CCIR notation, and the formula for relating $\Delta f/f$ to $\Delta t/T$ works. This is hardly an exact proof nor is the calculus that follows strictly rigorous, but it will serve for our purposes.

3.6 A MATHEMATICAL DERIVATION

In order to use time measurements to get frequency information, we need to know the relationships between time and frequency. A mathematical definition of frequency is:

$$f = \frac{1}{t},$$

where t is the period of the signal in seconds and f is the frequency in hertz. This can also be written as:

$$f = t^{-1}.$$

If we differentiate the frequency expression with respect to time and substitute we get:

$$df = -t^{-2}dt = -\frac{dt}{t^2} = -\frac{dt}{t} f$$

and

$$\frac{df}{f} = -\frac{dt}{t},$$

which can be written, for small changes, as:

$$\frac{\Delta f}{f} = -\frac{\Delta t}{t}.$$

31

Both sides of this equation are dimensionless. $\Delta f/f$ is, in the CCIR notation, F. Δt is the change in time read over a measurement time, T.

3.7 <u>DEFINITIONS</u>

f = actual frequency of source in hertz.

Δf = frequency difference in hertz.

f_n = nominal frequency in hertz.

F = relative frequency (dimensionless).

S = relative frequency difference (dimensionless)

CHAPTER 4.   USING TIME AND FREQUENCY IN THE LABORATORY

Although simple devices for the measurement of electrical frequency were invented during the 1800's, modern instruments have only been around since the middle of the twentieth century. Since World War II, however, time and frequency measurement has advanced at a rapid rate. Measurement of time intervals as small as one ten-billionth (10^{-10}) of a second and frequencies as high as one-hundred trillion (10^{14}) hertz (cycles per second) are now possible.

Such improvements, of course, are in keeping with the need for higher precision and accuracy in today's technology. Not long ago, the highest precision required for timing measurements was on the order of a millisecond. Today, however, the need for precisions of a microsecond is commonplace. For a variety of laboratory measurements, timing accuracies to the nearest nanosecond are required.

Time interval and its reciprocal, frequency, can be measured with greater precision than any other known physical quantity. By allowing two oscillators to run for a sufficiently long time, we can compare their average frequencies to a high degree of precision. Conversely, we can measure time intervals with the same precision by counting individual cycles in the output waveform of an oscillator, provided its frequency remains constant. But a problem arises in attempting to measure time intervals accurately.

Unlike two standard meter bars which can be left side by side indefinitely while their lengths are compared, time doesn't stand still for measurement. No two time intervals can be placed alongside each other and held stationary, so we must rely on the stability of an oscillator, a clock, a counter, or a similar measuring device to measure or compare time intervals.

In following the practices set forth in this chapter, one should keep in mind that the end result of a measurement can be no better than the performance of the equipment with which the measurement is made. Frequent and careful calibration of the measuring instruments against reliable standards is absolutely necessary if maximum accuracy is to be achieved.

In this chapter, we discuss the equipment needed to make time and frequency measurements in the lab, what the equipment is used for, how it is used, and what the results of using it will be. We do not discuss the reference sources of time and frequency signals being measured since they are fully discussed in later chapters.

4.1 ELECTRONIC COUNTERS

If you want to measure frequency or time interval, the modern electronic counter is a good investment. It is easy to use, does a large variety of measurements, and is rapidly going down in cost. The availability of new integrated circuits and displays has driven counter prices down while pushing their specifications upward. A 50 MHz counter is only a few hundred dollars or less at this writing. There is another advantage in using counters. This is the written material available on their construction and use. Articles in Ham radio publications, electronic hobby magazines, and manufacturers' application notes and other literature are plentiful, easy to find, and easy to read. In this chapter, only part of the available information is presented in a tutorial way. The reader is urged to obtain and study other material.

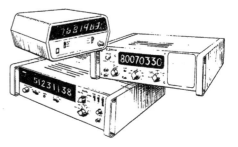

FIGURE 4.1.   ELECTRONIC COUNTERS.

Modern general-purpose counters have a variety of measurement features. Some have provisions for the insertion of plug-in modules which adapt the counter to an even greater range of applications. Although counters differ widely in specification and design details, all have several major functional sections in common. The basic parts of a typical counter are the time base, the main gate, and the decade counting assembly (DCA).

The time base provides uniformly spaced pulses for counting and control of the circuitry within the counter. The time interval between pulses must be accurately known because errors in the time base will appear directly in frequency and time measurements. Most counters incorporate an internal quartz crystal oscillator for a time base; some use the power-line frequency.

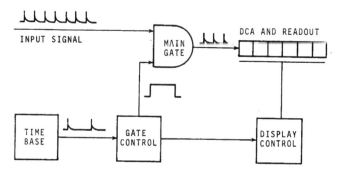

FIGURE 4.2. SIMPLIFIED BLOCK DIAGRAM OF AN ELECTRONIC COUNTER.

The oscillator may be followed by decade dividers to allow the user a choice of several different time signals. By adding more dividers to the chain, it is possible to obtain longer time intervals. By increasing the crystal oscillator frequency, one may obtain short time increments.

The main gate controls the time at which the count begins and ends. The gate may be actuated automatically by pulses from the time base or manually by means of a switch on the control panel. Pulses which pass through the gate are routed to the decade counting assembly where they are displayed on the visual numerical readout. After a pre-set display period, the instrument resets itself and starts another count.

Other counter sections include amplifiers, pulse shapers, power supplies, and control circuits for the various logic circuits. The input channels are usually provided with level controls and attenuators to set the exact amplitude limits at which the counter responds to input signals. Proper setting of the level controls allows accurate measurement of complex waveforms and also helps reject noise or other unwanted signals.

4.1.1 SOURCES OF COUNTER ERROR

The errors associated with using counters vary with the type of measurement. The most important source of error is the counter time base. This is the internal oscillator that controls the counting time or is itself counted and displayed. For accurate measurements, a good time base is essential. Often, the difference in cost of one counter over another depends on the quality of this internal oscillator. You get what you pay for, so study the time base specifications carefully when buying a counter. Errors are discussed more fully later in this chapter.

The second large source of error in counters is the ± 1 count ambiguity. Notice in figure 4.2 that the counter gate opens and closes to let the pulses through. Since the time base that drives the gate is not related in any way to the incoming pulses, sooner or later it is going to try to close or open the gate at a time when the pulse to be counted is already present. So the lack of synchronism between the signal counted and the gate will result in an extra count sometimes and one less count at other times. The effect on the counter display is the ± 1 count ambiguity. If the correct answer is 100.8, you will see the counter display either 100 or 101 some of the time. There are techniques for getting around this problem. Such techniques are discussed in counter manuals and also in manufacturers' application notes. Although one extra count out of ten million is often not significant, that same one count with a display of 100 could be very important.

The third main source of error using a digital counter stems from noise on the signals or from setting the controls. It is easy to be lulled into a sense of false security by the fact that the counter is displaying bright digits. But they may be wrong unless care is exercised in setting the controls so you know what signal is going into the counter. Accurate and precise measurements demand care and attention. Read the instruction manual carefully and know what it is you are measuring. Manufacturers often recommend the use of an oscilloscope in parallel with a counter. That way, you can see what is going on. Errors in counter measurement are discussed more fully later in this chapter.

4.1.2 FREQUENCY MEASUREMENTS

Modern counters can measure frequencies up to several hundred megahertz directly. With special methods of measurement, the frequency

34

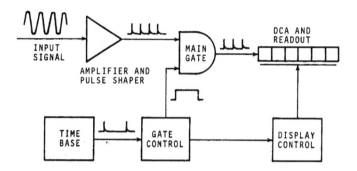

FIGURE 4.3. DIAGRAM OF A COUNTER IN THE FREQUENCY MEASUREMENT MODE.

range of the counter may be extended into the microwave region. The maximum counting rate of the basic instrument is limited by the inability of its switching circuits to follow extremely fast input signals.

A. Direct Measurement

For direct frequency measurements, the input signal is first amplified and converted into a series of uniform pulses. The pulses are then applied to the main gate, which is opened for a known time interval by the time base (see fig. 4.3).

The gate time may be 1 microsecond, 1 millisecond, 1 second, 10 seconds, etc., as determined by the setting of the time base selector switch. During the gating interval, the signal pulses pass through the main gate and on to the DCA, where the pulses are actually counted. The count obtained in a given gate time measures the average input frequency for that interval. The frequency is displayed on the readout and held there until a new count comes along. The decimal point in the display is automatically positioned to give the correct readout.

B. Prescaling

The high frequency limit of the counter can be extended by a technique known as pre-scaling. This means dividing the frequency of the input signal by some known factor before the signal is gated into the counter. Pre-scalers are available commercially.

If the signal frequency is prescaled by a factor of D, the counter displays 1/D times the actual frequency of the signal. By simple multiplication, the counter reading may be converted to the true value of the input frequency.

As an example, let us suppose we wish to measure a frequency in the range between 200 and 300 MHz with a counter capable of measuring to only 50 MHz directly. By means of a scale-of-8 divider, we reduce the unknown frequency to a value within the capability of the counter. If the counter indicates a frequency of 35.180 MHz, the actual value of the frequency being measured is 8 x 35.180 MHz = 281.44 MHz. In principle, any number may be chosen for the divider D, but presca-lers are usually designed to divide the input frequency by an integral power of two. This is because high speed counting circuits are binary; i.e., they divide by two. Integrated circuits are becoming available which will accept signals in the 100 MHz range and divide by ten.

To avoid the necessity of multiplying by the factor D, some prescalers incorporate a provision for lengthening the counter's gate time by the same factor. If the gating inter-val is increased D times while the input fre-quency is divided by D, the scale factor be-comes self-cancelling and the counter displays the correct input frequency. Figure 4.4 de-picts a counter equipped with a prescaler which divides the signal frequency and the time-base frequency by the same amount.

Prescaling offers a convenient way of covering wide frequency ranges without elabor-ate equipment and/or tedious operating proced-ures. Accuracy of the prescaler-counter com-bination is the same as that of the basic counter alone; but since the technique re-quires an increase in measurement time by virtue of the scale factor D, it is slower than the direct method.

C. Heteordyne Converters

Another way of extending the upper fre-quency limit of a counter is to use a hetero-dyne converter. The converter translates the

35

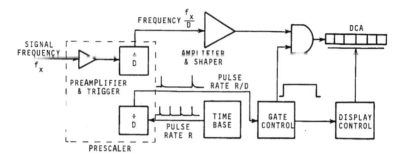

FIGURE 4.4. PRESCALER-COUNTER COMBINATION.

unknown signal frequency downward by beating it against a known signal of different frequency. The principle is the same as that employed in the familiar super-heterodyne radio receiver to produce an intermediate frequency. Heterodyne converters are available either as plug-in modules or as external accessories for many commercial counters. The block diagram of a typical, manually-tuned converter is shown in figure 4.5. The beat, or difference frequency, is applied to the counter and measured. The counter reading is added to the known signal frequency, and the sum is the unknown signal frequency.

Looking at figure 4.5, a 10 MHz frequency from the counter's internal time base is multiplied to 50 MHz and drives a harmonic generator. The harmonic generator produces a large number of frequencies, each 50 MHz apart. By means of a tunable cavity, the desired Nth harmonic can be selected from among the many available. In practice, it is best to pick the harmonic that is nearest to

but lower than the signal frequency. The mixer is then within the passband of the video amplifier as indicated by a movement of the tuning meter. The mixer output [f_x - (N x 50 MHz)] goes to the counter input and is measured. The dial marking on the cavity tuning control is added to the counter reading.

For example: Consider an input signal at 873 MHz. As the harmonic selector is tuned upward, higher harmonics of 50 MHz are injected into the mixer. When the tuning dial reaches 850 MHz (the 17th harmonic of 50 MHz), the tuning meter will rise sharply and the counter will display a reading of 23 MHz. Thus, the signal frequency is the sum of the dial setting (850) and the counter reading (23) or 873 MHz.

If the dial setting and the counter reading are to be added, the tuning point must always be approached from the lower side and stopped at the first peak. If the tuning control is turned too far, so that a harmonic

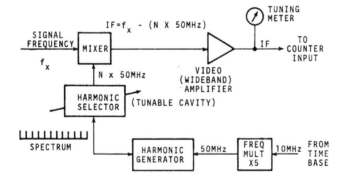

FIGURE 4.5. TYPICAL MANUALLY-TUNED HETERODYNE CONVERTER.

36

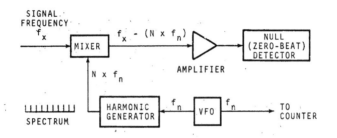

FIGURE 4.6. TYPICAL MANUALLY-TUNED TRANSFER OSCILLATOR.

higher than the input is selected, the counter reading must be treated as a negative number. Equipment operating manuals recommend that each measurement be repeated to be sure the answer is right.

The heterodyne technique offers nominally the same accuracy as direct counting. Of all known frequency-extension methods, the heterodyne technique allows the operator to make accurate microwave measurements in the least time. Also, this method can stand large amounts of frequency modulation on the measured signal without changing the accuracy of the measurement.

D. Transfer Oscillators

Transfer oscillators can be used to extend the frequency measuring range of counters beyond the limit of prescalers and heterodyne converters. Transfer oscillators are usable at frequencies up to 40 GHz and higher. Their wide bandwidth allows a single plug-in unit to cover a great range of frequencies. Also, most versions allow accurate frequency measurement of complex waveforms, such as frequency-modulated signals and pulsed rf.

A typical transfer oscillator consists of a stable variable frequency oscillator (VFO), a harmonic generator, a mixer, and a null detector. Like the heterodyne converter, the transfer oscillator mixes the unknown signal frequency with a selected harmonic of some internal reference frequency. But the reference frequency is derived from the VFO, and the counter is used (sometimes in conjunction with a prescaler or converter) to measure the VFO fundamental.

In operation, the VFO is carefully tuned until one of its harmonics matches the frequency of the signal being measured. The frequency match is indicated by a zero beat on the null detector. At this point the counter

reading f_1 is recorded. The VFO is then slowly tuned upward or downward until the next adjacent harmonic produces a null. A new counter reading f_2 is noted. The harmonic number applicable to f_1 can be calculated from the formula

$$N_1 = \frac{f_2}{|f_1 - f_2|} .$$

The bars mean a negative sign is ignored; i.e., the difference between f_1 and f_2 is always a positive number.

Having found both N_1 and f_1, the signal frequency, $f_x = N_1 f_1$. To illustrate this, let us assume that the first null occurs when $f_1 = 88.0$ MHz. As the VFO is tuned slowly upward, the next null is found at $f_2 = 90.0$ MHz. The applicable harmonic number is

$$N_1 = \frac{90.0}{|88.0 - 90.0|} = \frac{90.0}{2.0} = 45$$

and the unknown frequency is

$$f_x = 45 \times 88.0 \text{ MHz} = 3960 \text{ MHz.}$$

The same result would be obtained if the roles of f_1 and f_2 were interchanged; i.e., if the VFO were tuned downward from 88.0 MHz to 80.0 MHz in arriving at the second null.

For any given f_x, many null points may be noted as the VFO is tuned through its entire range. Using 3960 MHz as an example, an adjacent null pair may be found at VFO settings of $f_1 = 84.25531914...$MHz and $f_2 = 82.5$ MHz. Because the fractional part of f_1 does not end, however, the counter cannot display that number accurately. $N_1 = 47$ can be computed from the formula (rounded off to the nearest whole number since harmonic numbers must be integers). But f_x cannot be calculated with accuracy from the relation $f_x = N_1 f_1$ because there is no way of knowing the missing part of f_1. In this case, it is better to calculate f_x from f_2 since f_2 is an exact number.

An error may arise if the operator cannot tell exactly where the zero beat occurs. Here the resolution of the null detector is all important. The total reading error is the sum of the error in the counter reading plus the error caused by imperfect null detection. Typically, an uncertainty of several parts in 10^7 applies to microwave measurements made with manually-tuned oscillators.

Some of the more elaborate transfer oscillators employ feedback methods which lock the VFO to an exact submultiple of f_x. The phase lock feature essentially eliminates human error caused by improper zero-beat detection. Most automatic transfer oscillators also have provisions for extending the counter's gating interval by the factor N so that the counter can display f_x directly. Figure 4.7 shows block diagrams of three automatic transfer oscillators which are presently available commercially.

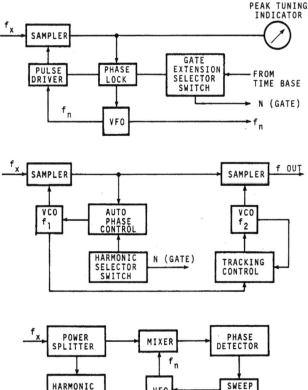

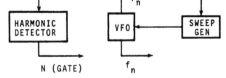

FIGURE 4.7. AUTOMATIC TRANSFER OSCILLATORS.

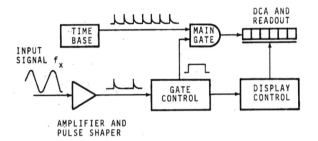

FIGURE 4.8. DIAGRAM OF COUNTER IN THE PERIOD-MEASUREMENT MODE.

Phase-locked transfer oscillators are normally used to measure only stable cw (continuous wave) signals and often are not applicable to complex waveforms containing frequency or pulse modulation. Even in those versions which are able to measure pulsed carriers, the accuracy is not as great as when measuring continuous waves.

4.1.3 PERIOD MEASUREMENTS

The frequency of a signal can be calculated from its period. Sometimes it is easier and better to measure the period of a signal rather than its frequency. The basic circuits of a frequency counter can be rearranged to make period measurements simply by exchanging the roles of the time base and the input signal (fig. 4.8). This is done by means of a switch on the control panel. For period measurements, the output of the time base generator is counted during a time interval controlled by the period of the input signal f_x.

If, for example, f_x has a period of 1 ms and the time base output is 10^5 pps, the counter displays $(1 \times 10^{-3} \text{ s})(1 \times 10^5 \text{ pulses/s}) = 100$ pulses during each gate interval. If

the period of f_x is increased to 10 ms, the counter totals 1000 pulses. The counter readout is proportional to the period of f_x and can be designed to display the period directly in units of time (microseconds, milliseconds, seconds, etc.).

Frequency obtained by period measurements is often more accurate than direct frequency measurements. This is true up to the point where the frequency being measured equals the frequency of the counter time base oscillator. The disadvantage of period measurements is in interpreting the meaning of the counter display. It is not hertz or cycles per second-- it is time. To get hertz you have to divide the answer into the number 1 (take the reciprocal). But this disadvantage is rapidly disappearing due to the new integrated circuits now available. The division is performed by the counter circuits and you get the best of both worlds--period measurements for accuracy with the answer displayed in hertz.

The accuracy of a period measurement can be improved further by using a multiple-period mode of operation. In this mode, the output of the signal amplifier is routed through one or more decade frequency dividers before

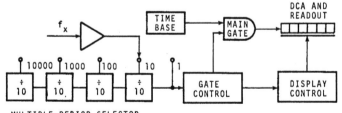

MULTIPLE-PERIOD SELECTOR

FIGURE 4.9. DIAGRAM OF COUNTER IN THE MULTIPLE-PERIOD MODE.

PERIOD MULTIPLE	TIME BASE				
	1 ms	0.1 ms	10 μs	1 μs	0.1 μs
X 1	2 ms	2.4 ms	2.40 μs	2408 μs	2409.0 μs
X 10	2.4 ms	2.41 ms	2409 μs	2409.0 μs	2408.91 μs
X 100	2.41 ms	2.408 ms	2408.8 μs	2408.91 μs	2408.912 μs
X 1,000	2.409 ms	2.4089 ms	2408.91 μs	2408.911 μs	2408.9113 μs
X 10,000	2.4089 ms	2.40890 ms	2408.912 μs	2408.9112 μs	*408.91124 μs

*COUNTER HAS ONLY AN 8-DIGIT DISPLAY

reaching the gate control circuit. Thus the main gate is held open for 10, 100, 1000, or more periods depending upon the number of divider stages used. The more periods over which a signal is averaged, the better the resolution. But you have to wait longer for the answer.

To illustrate the advantage of multiple-period averaging, consider a counter reading of 415 kHz taken in the frequency-measurement mode. The same input, if divided by 10 using a period measurement with a 10 MHz time base would display a six-digit reading of 2408.91 microseconds. By taking the reciprocal, f_x averages 415.125 Hz over ten complete periods. We gained three decimal points by period averaging.

The accuracy of the measurement may be improved in two ways: by increasing the time base frequency (a more costly counter) or by averaging for more periods (waiting longer). Table 4.1 shows the results of repeated period measurements of the same signal. The measurements were made with a counter having an 8-digit display and with various settings of both the time base selector switch and the period multiplier switch. The improved resolution at higher time base rates and at longer gating intervals is quite apparent.

Noise is a major source of error in period measurements. A noisy input signal may cause the main gate to open or close too soon at one time and too late at another. In the period measurement, assume that the counter is designed to trigger the main gate at a positive-going zero-axis crossing of the input signal. Any jitter at this point produces erratic gating, as shown in figure 4.10.

The error resulting from noise is independent of frequency and may be reduced by extending the measurement to more than one period. A ten-period measurement, for instance, is ten times more precise than a single-period measurement because the same uncertainty in the time of opening and closing the main gate is distributed over an interval ten times longer.

4.1.4 TIME INTERVAL MEASUREMENTS

Electronic counters are widely used for the measurement of time intervals, both periodic and aperiodic. They range in length from less than a microsecond up to hours and even days. At the end of the measured interval, the elapsed time is displayed in digital form.

Counters vary considerably in their time-measuring ability. Some are designed to measure the duration of a single electric pulse while others measure the interval between th occurrence of two different pulses. The mor versatile models have individually adjustabl start and stop controls which allow greate control by the operator.

As with period measurements, the basi components of the counter are connected i such a way that the output pulses from th time base are counted. To select the interva being measured, separate start-stop amplifier and controls are used to open and close th main gate.

40

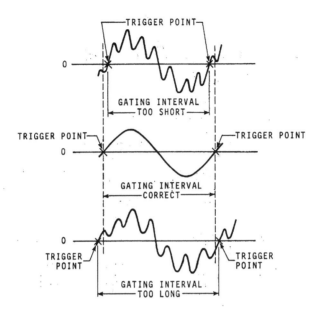

FIGURE 4.10. EFFECT OF NOISE ON TRIGGER POINT IN PERIOD MEASUREMENTS.

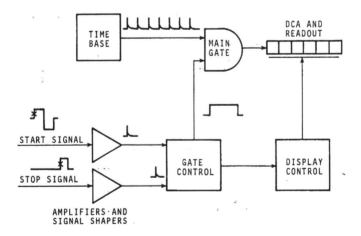

FIGURE 4.11. DIAGRAM OF COUNTER IN THE TIME INTERVAL MODE.

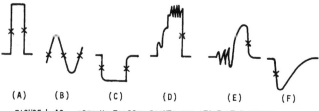

(A)   (B)   (C)   (D)   (E)   (F)

FIGURE 4.12.   OPTIMUM TRIGGER POINTS FOR START-STOP PULSES.

During the time between the start and stop signals, the main gate is held open and counts are accumulated according to the length of the gate interval. At the end of the gate interval, the readout assembly can be made to display the elapsed time directly in microseconds, seconds, minutes, or other appropriate units.

The start-stop input circuits are designed to produce sharp, uniform trigger pulses from a variety of signals. Best results are obtained if clean pulses are used. If the instrument has a trigger level control, it is possible to select the exact input amplitude at which a trigger pulse from that channel is generated. If a slope control is provided, it is possible to set the trigger point on either the positive-going or the negative-going side of the input signal at will.

Careful setting of the trigger level and slope controls is necessary to avoid measurement errors caused by noise, jitter, harmonics, and other forms of distortion on the input signal. Figure 4.12 shows the best triggering points for six different input waveshapes. In each case, the desired trigger point (as indicated by an x) is at the center of the steepest, cleanest portion of the signal. Note that for noise-free sinusoidal (sine wave) signals as in waveform (B), the optimum trigger points occur at the zero crossings.

4.1.5 PHASE MEASUREMENTS

The time interval counter can be used to measure phase relationships between two input signals of the same frequency. To do this, one of the signals goes to the start channel while the second signal goes to the stop channel. The level and slope controls are adjusted to trigger at the same point on both input waveforms. For sinusoidal waves, the best triggering point is the zero-crossover. The time interval between corresponding points on the two input waveforms is then read directly from the counter. Time interval can be converted to a phase angle by the formula

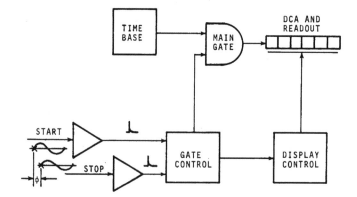

FIGURE 4.13.   USE OF TIME-INTERVAL UNIT FOR PHASE MEASUREMENTS.

42

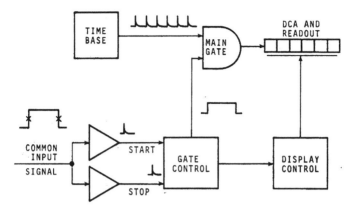

FIGURE 4.14. TIME-INTERVAL COUNTER WITH START-STOP CHANNELS CONNECTED TO COMMON SOURCE FOR PULSE-WIDTH OR PERIOD MEASUREMENTS.

$$\phi \text{ (PHASE ANGLE)} = \frac{360 \times \text{Time Interval}}{T} \text{ degrees}$$

$$= \frac{2\pi \times \text{Time Interval}}{T} \text{ radians,}$$

where T is the known period of the two signals expressed in the same units as the time interval.

4.1.6 PULSE-WIDTH DETERMINATION MEASUREMENTS

Trigger controls are seldom calibrated with sufficient accuracy to measure the rise time or decay time of a single pulse. It is almost always necessary to use an oscilloscope to measure rise time. An oscilloscope used with a counter is very useful. By selecting one polarity of slope for the start trigger and the opposite polarity for the stop trigger, you can measure the total duration of a pulse. The duration of pulse (A) in figure 4.12, for example, could be determined by selecting point x on the positive slope for the start trigger and point x on the negative slope for the stop trigger. Some time interval counters bring the gate start-stop pulse out on connectors for oscilloscope viewing.

Carrying the idea one step further, the period of waveform (B) in figure 4.12 could be measured by choosing the positive zero-crossing as the trigger point for both the start and stop pulses. When measuring pulse width or period with a time interval counter, it is necessary to connect both input channels to the same source (figure 4.14).

The smallest measurable time interval is limited by the resolution of the counter. In general, the resolution of time interval and period measurements are affected by the same factors. The presence of noise can seriously impair the precision of pulse-width determinations.

4.1.7 COUNTER ACCURACY

Depending upon the characteristics of the counter and its method of use, the accuracy of measurements may range from 1 part in 10^2 to a few parts in 10^{10} or better. Although the possible causes for inaccurate measurement are many, the main sources of error are (1) time-base instability, (2) uncertain gating, and (3) faulty triggering. The total measurement error is the algebraic sum of all individual errors.

A. Time Base Error

The time base oscillator will change frequency slightly during the course of a measurement. This is due to temperature changes, line voltage changes, internal noise, and crystal aging. Temperature effects can be minimized by allowing the instrument to warm up before use and by keeping the counter in a constant-temperature location. If the line voltage fluctuations are severe, it may be necessary to operate the counter from a regulated power source. Check the operating manual for the manufacturer's recommendations.

43

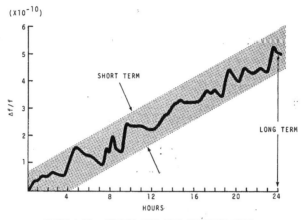

$(\times 10^{-10})$

HOURS

FIGURE 4.15. TYPICAL TIME BASE STABILITY CURVE.

Any noise generated internally by the time base oscillator produces a jitter in its output frequency. This may cause erroneous counts, especially when a measurement is performed during a brief time interval. As the measurement time is made longer, the jitter tends to average out. For this reason, the jitter is often called "short-term" instability. A typical time base stability graph is shown in figure 4.15. Note that the vertical axis is labeled $\Delta f/f$, which is equal to F, the relative frequency of the time base. A complete explanation of this notation is given in Chapter 3.

In modern counters, the amount of oscillator noise is generally so small that short-term stability is specified only for the most precise instruments. When reading specifications, it should be realized that the averaging time has an important bearing on the actual meaning of stability figures. For example, an oscillator having a stability specification of 5 x 10^{-10} (per minute) might be noisier than one with a specification of 5 x 10^{-9} (per second). To be absolutely meaningful, short-term stability figures should be compared for the same averaging period, preferably of only a few seconds.

Long-term instability, on the other hand, refers to a slow, often predictable drift in the average frequency of an oscillator due to gradual changes in the crystal. Quartz crystals exhibit drift rates typically of 10^{-10} to 10^{-6} per day as a result of crystal aging. After a brief warmup period when the oscillator is first turned on, a high-quality crystal has a linear frequency drift, the slope of which reveals the crystal aging rate.

Because aging produces larger and larger error as time goes on, it is necessary to recalibrate the oscillator if the time base is to be accurate. For utmost accuracy it may be wise to check the oscillator calibration immediately before and after a critical measurement. However, on inexpensive counters, this may be impractical.

For measurements which require a better time base accuracy than that of the counter's internal oscillator, you can use an external frequency source to drive the time base generator. Most counters have a built-in connector and switch for using an external oscillator. The internal oscillator is automatically disconnected when the time base is switched to an external source.

B. Gate Error

As previously mentioned, an uncertainty of ± 1 count is inherent with all conventional electronic counters because the time base pulses are usually not synchronized with the signal being measured. Because of this, it is possible for the reading to be in error by one count even if there are no other sources of error! As an example, figure 4.16 shows how the same signal might produce a count of either 8 or 9, depending upon when the gating interval started.

It is obvious that the percentage error becomes smaller as more pulses are counted. A one count matters less for 1,000,000 than for 1,000. This explains why long gating intervals permit greater accuracy of measurement.

44

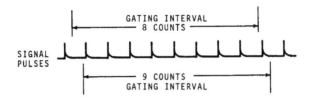

FIGURE 4.16. CONSTANT GATING INTERVAL WITH AMBIGUITY OF ± 1 COUNT.

If we count a 1 MHz frequency for 1 second, the ± 1 count error is one part per million. If either the measured frequency is increased to 10 MHz or the gating interval is lengthened to 10 seconds, the error goes down to 1 part per ten million. In measurements of high frequencies, made directly or with the aid of heterodyne converters, the uncertainty of ± 1 count may be less important than the counter's time base error. The same is true of prescalers which incorporate gate extension features.

Previously, we noted that low frequencies can be measured more accurately by period measurements than by direct frequency measurements. For very high frequencies, however, the reverse is true. At some point, the effect of gate error must be identical for either mode of operation. The chart in figure 4.17 depicts the crossover point as well as the accuracy which can be expected both from multiple-period measurements and direct frequency measurements up to 100 MHz.

C. Trigger Errors

Trigger errors usually arise from the presence of noise or other modulation on the gate-control signal. This source of error has been mentioned previously in conjunction with period and time interval measurements. For frequency measurements, the trigger error is almost always negligible because the gate-control signal, obtained directly from the time base, can be virtually noise-free.

For period and time interval measurements, the trigger error (ε_{tr}) is expressed mathematically as the ratio of the gate time deviation (Δt) to the total duration (T) of the gating interval. Trigger error is also proportional to the fraction of peak noise voltage (E_n) over peak signal voltage (E_s).

$$\varepsilon_{tr} = \frac{\Delta t}{T} = k \frac{E_n}{E_s},$$

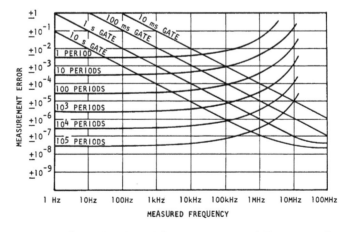

FIGURE 4.17. ACCURACY CHART FOR PERIOD AND FREQUENCY MEASUREMENTS.

45

where k is a proportionality constant determined by the nature of the modulation.

Although noise can sometimes be removed from the input signal by external devices; e.g., limiters, filters, etc., its effects can be reduced more conveniently by multiple-period averaging. As more periods are averaged, the effects of both the ± 1 count ambiguity and the trigger error diminish proportionately.

Most counters incorporate a Schmitt trigger or similar circuit to generate the trigger pulses. Because of a slight hysteresis or lag in switching action which characterizes such circuits, the rise and decay of the trigger pulses are not started at exactly the same voltage. Instead, the width of the trigger pulse is determined by two voltage levels--one that establishes the point at which the pulse

begins to rise and the other that dictates the point where the pulse starts to fall. A complete pulse is produced, therefore, only when the input signal crosses both critical levels.

Figure 4.18 shows how counts may be lost or gained as a result of signal modulation. The shaded area in each example is the hysteresis zone of the trigger circuit. Incorrect counts are likely to result whenever the input waveform reverses direction within the shaded zone.

To measure the time interval between the occurrence of a sharp pulse and the next positive-going zero crossing of a continuous sine wave, it would be logical to start the counter with the pulse and adjust the stop-channel control(s) to trigger at zero volts on the positive slope. If the trigger level control for the stop channel is improperly

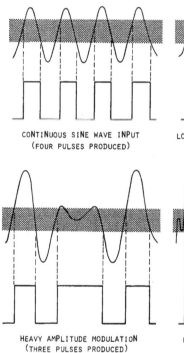

CONTINUOUS SINE WAVE INPUT
(FOUR PULSES PRODUCED)

LOW-FREQUENCY RIPPLE SUPERIMPOSED
(THREE PULSES PRODUCED)

HEAVY AMPLITUDE MODULATION
(THREE PULSES PRODUCED)

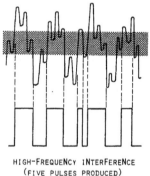

HIGH-FREQUENCY INTERFERENCE
(FIVE PULSES PRODUCED)

FIGURE 4.18. UNDIFFERENTIATED SCHMITT-TRIGGER WAVEFORMS.

46

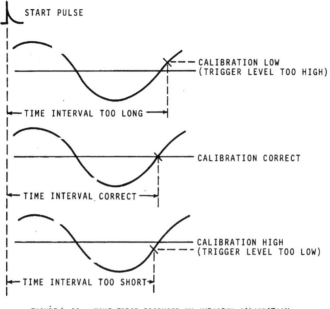

START PULSE

CALIBRATION LOW
(TRIGGER LEVEL TOO HIGH)

←——TIME INTERVAL TOO LONG——→

CALIBRATION CORRECT

←——TIME INTERVAL CORRECT——→

CALIBRATION HIGH
(TRIGGER LEVEL TOO LOW)

←——TIME INTERVAL TOO SHORT→

FIGURE 4.19.   TIME ERROR PRODUCED BY IMPROPER CALIBRATION
OF TRIGGER LEVEL CONTROL.

calibrated, however, the stop pulse may be produced at a point that is actually positive or negative when the control is set at zero. This means that the measured time interval will be too long or too short as shown by figure 4.19.

Of course, an opposite error will result if the trigger level control for the start channel reads too low or too high. If the controls for both channels are misadjusted in opposite directions, the combined error will be cumulative. Furthermore, the effect of trigger calibration error is made worse if instead of a steep wavefront the input signal has a gradual slope at the selected triggering point.

A low-frequency sine wave is convenient for checking the calibration of trigger controls. With the counter in the time interval mode, the gating interval switch in the X1 position, and both input channels adjusted to trigger at zero volts on the same slope, the counter should display exactly one period of the test signal. Next the slope selector for the stop channel is turned to its opposite position without adjusting any other controls. If the counter reading changes by exactly one-half period when the opposite slope is

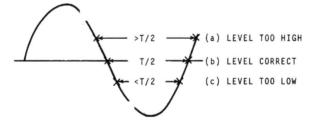

>T/2 ———— (a) LEVEL TOO HIGH

T/2 ———— (b) LEVEL CORRECT

<T/2 —→ (c) LEVEL TOO LOW

FIGURE 4.20.   SINE WAVE METHOD OF CHECKING TRIGGER LEVEL CALIBRATION.

| DIGITAL COUNTER | ---- | D TO A CONVERTER |

| S |
| ANALOG CHART STRIP RECORDER |

FIGURE 4.21. DIGITAL-TO-ANALOG ARRANGEMENT FOR CHART RECORDING.

selected, the trigger level setting is correct. If the reading changes by more than one-half period, as in (a) of figure 4.20, the stop channel's trigger level is too high. If the change is less than one-half period, as in (c), the trigger level is too low. The start trigger can be checked in a similar fashion by holding all other controls fixed and switching the slope selector for the start channel only.

4.1.8 PRINTOUT AND RECORDING

Some counters have provisions for connecting a printer to get a permanent record of the readings. Using a digital-to-analog converter, the counter can also be used to drive a chart recorder. The printer must be compatible with the counter if a valid printout is to be obtained.

4.2 OSCILLOSCOPE METHODS

The cathode-ray oscilloscope is probably one of the most useful laboratory tools available to the electronics technician, engineer, or scientist. Whole volumes have been devoted to its many uses.

The basic physical principles of the oscilloscope have been known for years. An early oscillograph (as it was then called) was in operation during the 1890's. This device used wires, mirrors, and light beams instead of electronic circuits, but it could display waveforms of low-frequency alternating current.

The modern cathode-ray tube was developed in the early 1920's. Since then, the electronic oscilloscope has evolved into an instrument that can display high-frequency waveforms and store images for days. Measurements or comparisons of frequencies and time intervals are among the major uses of the oscilloscope.

There are four main parts: (1) the cathode-ray tube and its power supply, (2) the vertical deflection amplifier, (3) the horizontal deflection amplifier, and (4) the time-base generator. A block diagram is shown in figure 4.22.

The role of the time-base generator is fairly involved. In order to display faithfully a signal that is a function of time, the spot of light must traverse the cathode-ray tube screen at a uniform rate from one side to the other and then return to the original side as quickly as possible. The voltage that causes this horizontal sweep is generated by the time-base generator. The waveform is a ramp or "sawtooth" as shown in figure 4.23.

The frequency of the ramp voltage is variable so that waveforms of any frequency

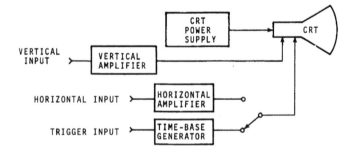

FIGURE 4.22. SIMPLIFIED BLOCK DIAGRAM OF A CATHODE-RAY OSCILLOSCOPE.

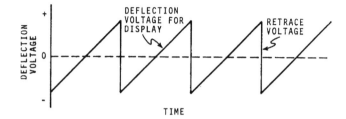

FIGURE 4.23. WAVEFORM OF HORIZONTAL DEFLECTION VOLTAGE FROM TIME-BASE GENERATOR.

within the limits of the oscilloscope can be displayed. The time-base generator has a control to change the frequency of the ramp. How fast the spot moves across the screen is of interest to the operator; therefore, the frequency control is usually labeled in seconds (or fractions of seconds) per centimeter.

When the frequency of the traverse is high enough, the persistence of the phosphorous and the retention capability of the human eye combine to make the moving spot appear as a solid line across the screen. If a periodic wave is applied to the input of the vertical amplifier and the time-base generator is properly adjusted, the waveform will be displayed as a stationary pattern on the cathode-ray tube screen.

4.2.1 CALIBRATING THE OSCILLOSCOPE TIME BASE

If there is much doubt as to the accuracy of the time-base generator, it should be calibrated. Use a signal that is more accurate than the time-base signal being measured. The calibration is done in much the same way as the frequency measurement, but in reverse. If you have a crystal oscillator that operates in the same frequency range as the sweep time base being measured, proceed as follows:

Display the calibration source signal on the screen, and adjust the time-base switch to the setting that should give one cycle per centimeter. If the source is 100 kHz, for example, the time-base step switch should be set to 10 microseconds/centimeter. If the time-base generator is off frequency, there will be more or less than one cycle per centimeter displayed on the screen. Adjust the variable frequency control of the time-base generator until the display is correct.

The adjustment may or may not hold for other positions of the step switch, so it is best to recalibrate each position. Many laboratory oscilloscopes have built-in calibration oscillators which can be used in the same way as an external frequency source. Refer to the oscilloscope manual for details.

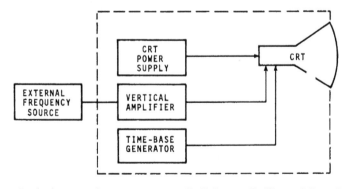

FIGURE 4.24. TIME BASE CALIBRATION HOOKUP USING AN EXTERNAL FREQUENCY SOURCE.

49

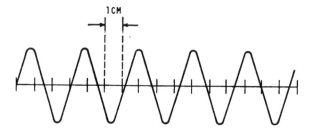

FIGURE 4.25. SINE WAVE DISPLAY AS VIEWED ON OSCILLOSCOPE.

4.2.2 DIRECT MEASUREMENT OF FREQUENCY

Suppose that we want to measure the frequency of a sine wave when it appears as shown in figure 4.25. By counting the number of cycles and the number of scale divisions (centimeters), we see that five cycles occur in 15.5 centimeters. Now we observe the time-base generator dial, first making sure it is in the calibrated mode. If the dial reads 0.1 millisecond per centimeter, for instance, the frequency computation would be:

$$\frac{5 \text{ cycles}}{15.5 \text{ cm}} \times \frac{1 \text{ cm}}{0.1 \text{ ms}} \times \frac{1000 \text{ ms}}{1 \text{ s}} \cong 3230 \text{ Hz}.$$

Alternatively, we may divide the display (in cycles per centimeter) by the time-base rate (in seconds per centimeter):

$$\frac{\text{cycles}}{\text{cm}} \div \frac{\text{s}}{\text{cm}} = \frac{\text{cycles}}{\text{cm}} \times \frac{\text{cm}}{\text{s}} = \text{Hz}.$$

Direct frequency measurements with the oscilloscope have an accuracy limit of about 5 parts per thousand. Accuracy of the measurement is limited by the ability of the operator to read the screen and by the accuracy of the time base. In general, one should expect to do no better than about 0.5%. Still, if you want to know whether you are looking at 60 Hz or 120 Hz, or if you are checking a radio carrier frequency for harmonics, the accuracy is good enough.

4.2.3 FREQUENCY COMPARISONS

Unlike direct frequency measurements, frequency comparisons with the oscilloscope can easily be made with a precision of one part per million. A precision of one part per 100 million or even one part per billion can be achieved by using extra equipment described later.

A. Lissajous Patterns

The optical-mechanical oscillograph was developed in 1891, but the idea of a Lissajous pattern is even older, having been demonstrated by a French professor of that name in 1855.

If we apply a sine wave simultaneously to the horizontal and vertical inputs of an oscilloscope, the spot will move in some pattern that is a function of the voltages on each input. The simplest Lissajous pattern occurs if we connect the same sine wave to both inputs. The equipment hookup is shown in figure 4.26. The resulting pattern is a straight line. Furthermore, if both signals are of the same amplitude, the line will be inclined at 45° to the horizontal.

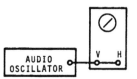

FIGURE 4.26. EQUIPMENT HOOKUP
FOR A 1:1 LISSAJOUS
PATTERN.

If the input signals are equal in amplitude and 90° out of phase, a circular pattern results. Other phase relationships produce an elliptical pattern. The phase difference in degrees between the horizontal sine wave and the vertical sine wave can be determined as shown in figure 4.27.

The phase relationship between signals that generate linear, circular, or elliptical patterns is given by:

$$\sin \theta = \frac{A}{B},$$

50

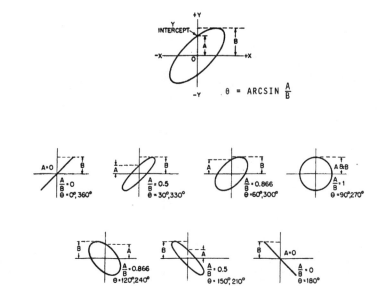

$$\theta = \text{ARCSIN } \frac{A}{B}$$

FIGURE 4.27. ELLIPTICAL LISSAJOUS PATTERNS FOR TWO INDENTICAL FREQUENCIES OF DIFFERENT PHASE.

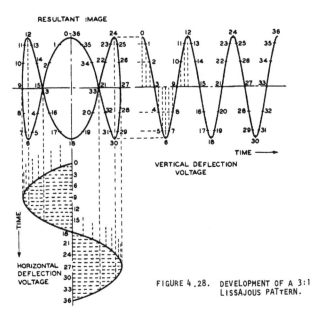

RESULTANT IMAGE

VERTICAL DEFLECTION VOLTAGE

TIME

HORIZONTAL DEFLECTION VOLTAGE

FIGURE 4.28. DEVELOPMENT OF A 3:1 LISSAJOUS PATTERN.

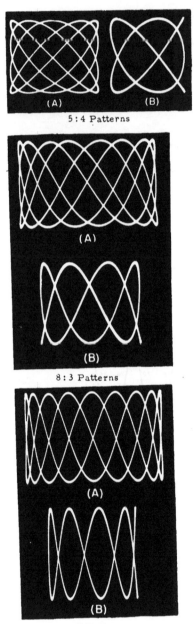

5:4 Patterns

8:3 Patterns

9:2 Patterns

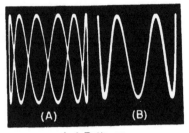

6:1 Patterns

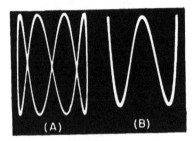

4:1 Patterns

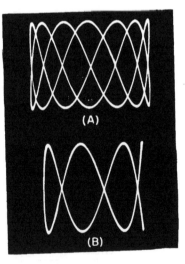

7:2 patterns

FIGURE 4.29. LISSAJOUS PATTERNS.

where θ is the angle between the two sine waves, and A and B are relative distances measured along the vertical axis of the trace.

The actual determination of frequency ratio is done in a straightforward manner. The number of loops touching, or tangent to, a horizontal line across the top of the pattern is counted. Likewise the number of loops tangent to either the right or left side is counted. The frequency ratio is then computed from the formula

$$\frac{f_v}{f_h} = \frac{T_h}{T_v} ,$$

where: f_v = the frequency of the vertical deflection voltage,

   f_h = the frequency of the horizontal deflection voltage,

   T_h = the number of loops tangent to the horizontal line, and

   T_v = the number of loops tangent to the vertical line.

Figure 4.28 shows the graphical construction of a 3:1 pattern. Note that the ratio of T_h/T_v is 3:1.

Lissajous patterns become quite complicated when the two frequencies are widely separated. Some oscillograms for higher ratios are shown in figure 4.29. Two versions of each pattern are shown. Version (A) is the typical closed-lobe pattern, and version (B) is the double image that results if phase relations cause the pattern to be folded and exactly superimposed upon itself.

The usefulness of Lissajous patterns depends upon a number of things. Consider a 1:1 pattern. If we are trying to synchronize two power generators at 60 Hz and our pattern is changing through one complete cycle every second, we have achieved synchronization to 1 Hz or about 1.7%--not very good. If our frequency was 100 kHz, the same change of one revolution per second in our pattern would give a resolution of 1×10^{-5}. This shows that the higher the frequencies we are working with, the better synchronization becomes.

In comparing or synchronizing precision oscillators, a frequency multiplier may be useful. Every time our signals are multiplied in frequency by a factor of ten, the precision of measurement improves by a factor of ten. When the two signals are multiplied and then fed to a phase detector, they produce a low-frequency signal equal to the difference in hertz between our two multiplied signals. This difference signal, Δf, may then be compared via Lissajous pattern against the sine wave from a precision low-frequency oscillator.

If we again use two 100-kHz signals as an example but multiply their frequencies to 100 MHz before comparing them, we achieve a measurement precision of 1×10^{-8}. This arrangement is shown in figure 4.30.

The phase detector may be a double-balanced mixer of the type shown in figure 4.31. These mixers are available commercially.

Another system for comparing two oscillators is indicated in figure 4.32. A phase-error multiplier increases the phase or frequency difference between the oscillators by some factor, say 1000, without frequency

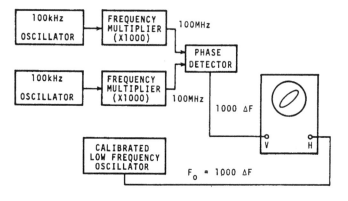

FIGURE 4.30.   ARRANGEMENT FOR MEASURING SMALL FREQUENCY DIFFERENCES WITH OSCILLOSCOPE AND FREQUENCY MULTIPLIERS.

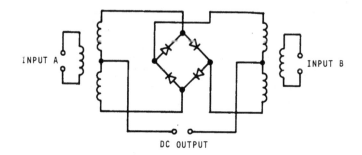

FIGURE 4.31. DOUBLE-BALANCED MIXER.

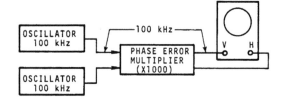

FIGURE 4.32. ARRANGEMENT FOR FREQUENCY COMPARISON USING
OSCILLOSCOPE AND PHASE-ERROR MULTIPLIER.

multiplication. The output from the phase-error multiplier can be connected directly to the oscilloscope. The resulting Lissajous pattern will allow oscillator synchronization with the same precision as provided by the frequency multiplier technique. If care is taken, calibrations with errors as small as 1×10^{-9} or 1×10^{-10} can be accomplished.

A principal use of the 1:1 Lissajous pattern is for synchronization of oscillator frequencies. If the pattern is drifting, no matter how slowly, it means that a frequency difference exists between the oscillators. Neither the oscilloscope nor the operator can tell which oscillator is at the higher frequency. The Lissajous pattern is the same regardless of which frequency is higher. What can be done, though, is to slightly change the frequency of one of the oscillators. If the rate of pattern drift increases when the frequency of oscillator #1 is raised, then oscillator #1 is at a higher frequency than oscillator #2. The opposite happens if the pattern drift slows down when the frequency of oscillator #1 is increased.

Calibration of variable oscillators such as rf signal generators and audio oscillators is easily done using Lissajous patterns of higher than 1:1 ratio. Figure 4.33 shows a

method of calibrating an rf signal generator using a fixed 100-Hz frequency source.

Starting with a 1:1 Lissajous pattern between the standard oscillator and the signal generator, mark the calibration points on the generator dial each time the pattern becomes stationary as the dial is turned. Each stationary pattern represents a definite frequency ratio which can be evaluated by one of the methods explained previously. Patterns of ratios up to 20:1 can be recognized if the signal generator is stable enough.

FIGURE 4.33. ARRANGEMENT FOR CALIBRATING
A SIGNAL GENERATOR.

54

If the oscillator has a base frequency of 1 MHz with internal dividers to 100 kHz, calibration points can be found on the signal generator dial every 100 kHz from 0.1 MHz to 2 MHz and every 1000 kHz from 2 MHz to 20 MHz. This range is adequate for many rf signal generators. Multipliers can be used to extend the markers into the VHF range if necessary, but the calibration points will be farther apart.

Not much can be learned from a drifting pattern of high ratio, and even the stationary pattern may be confusing. A complete rotation of the pattern in one second means a 1-Hz deviation by the higher frequency no matter what the ratio. But for ratios other than 1:1, the change in the lower frequency is not as great.

With a 10:1 pattern, for instance, a drift of one revolution per second could mean that the higher frequency is off by 1 Hz, or it could mean that the lower frequency is off by 0.1 Hz. Unless one of the two frequencies is assumed to be accurate, there is no way of telling from the Lissajous pattern which frequency is in error. Only their relative difference can be measured by this method (see Chapter 3).

B. Sweep Frequency Calibration: An Alternative Method

Most frequency comparisons made in a standards laboratory are between oscillators that have very small frequency differences. A 1:1 Lissajous pattern between oscillators having a relative frequency difference of 1×10^{-11} would take over a day if the comparison were made at 1 MHz! There are faster ways to measure the rate of change on the oscilloscope. These methods allow the difference frequency to be determined in much shorter times than is possible with a slow moving Lissajous pattern.

The output wave of an oscillator to be measured is displayed on the oscilloscope. Next, the oscilloscope sweep frequency oscillator is calibrated. Now, if the oscilloscope is triggered by a reference oscillator, the pattern or waveform will appear to stand still if the frequencies of the two oscillators are nearly equal. Actually, the pattern will slowly drift. The rate of drift in microseconds per day depends on the frequency difference between the oscillators. The direction of drift tells us which oscillator is high or low in frequency with respect to the other.

When a sine wave of an oscillator frequency is displayed and the oscilloscope is triggered by a reference oscillator, the movement of the pattern can be used for calibration. This pattern will move to the left or right if the unknown frequency is high or low with respect to the reference oscillator. The question is, how fast is the pattern moving? How many microseconds is it moving in how many hours or days? Remember, the sweep is calibrated in microseconds or milliseconds per centimeter.

One could make a measurement of the pattern position at, say, 10:05 am, then come back at 2:30 pm for another look. Suppose in the meantime, it moves exactly three centimeters and the sweep speed is set at one microsecond per centimeter. What does this tell us?

ELAPSED TIME = 2:30 pm - 10:05 am

$$= 4 \text{ hours } 25 \text{ minutes} = 0.184 \text{ day.}$$

$$\text{PHASE DRIFT RATE} = \frac{3.00 \text{ cm} \times 1.00 \text{ } \mu s/cm}{0.184 \text{ day}}$$

$$= 16.3 \text{ } \mu s/day.$$

This result tells us how much our oscillator moved in time (16.3 microseconds) during an interval of one day. The result is in the form $\Delta t/T$ (see Chapter 3 for a further explanation). Expressed as a numeric, $\Delta t/T = 1.89 \times 10^{-10}$ since there are $86,400 \times 10^6$ microseconds in a day.

We can relate this to a frequency notation because Chapter 3 tells us that

$$-\frac{\Delta t}{T} = \frac{\Delta f}{f} \quad .$$

For this oscillator, the relative frequency, $F = \Delta f/f = -1.89 \times 10^{-10}$. The negative sign tells us that the measured frequency was lower than its nominal value. In this case, it is assumed that the reference frequency is equal to its nominal value. Notice also that the oscilloscope has performed a measurement much like the time interval counter mentioned earlier in this chapter.

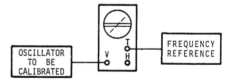

FIGURE 4.34. VIEWING FREQUENCY DRIFT.

4.2.4 TIME INTERVAL MEASUREMENTS

Time-interval measurements are quite easy to perform with the oscilloscope. Usually, one time pulse is displayed on the screen while a different time pulse is used to trigger the oscilloscope. Another method is to display both time pulses on a dual-trace scope. Triggering can be achieved from either of the two displayed pulses or from a third time pulse. An adjustable rate divider is handy for this purpose.

A point worth remembering is that the accuracy of the time-interval measurement is inversely proportional to the length of the interval. If the time pulses are 10 milli-seconds apart, for instance, the measurement can be performed to probably no better than ± 50 microseconds. But if the pulses are only 100 microseconds apart, the measurement can be performed with an accuracy of ± 0.5 microsecond. In both cases, however, the relative uncertainty is the same (± 5 parts in 10^3).

For time pulses spaced close together, the pulse width may occupy an appreciable percent of the time interval; therefore, large errors can be introduced by having the wrong trigger slope or improper trigger amplitude. Unless care is taken by the operator, such errors can nullify the accuracy that is other-wise attainable in the measurement of brief time intervals.

The main advantage in using an oscillo-scope for time interval measurements is that one can see the pulse waveform. For this reason, time intervals between odd-shaped or noisy pulses are better measured with the oscilloscope than with other types of instru-ments. Time intervals in the submicrosecond range can be measured easily and economically with high-frequency oscilloscopes, but long intervals are more accurately measured with an electronic counter, as described earlier. Always refer to your oscilloscope manual; it contains additional useful information.

4.3 WAVEMETERS

A wavemeter is an adjustable circuit that can be made to resonate over a range of fre-quencies as indicated on a calibrated dial. The dial may be marked in units of frequency or wavelength. A typical wavemeter covers a frequency range on the order of one octave. Resonance is detected by observing the re-sponse of a meter or other device incorporated into the measuring instrument or by noting a disturbance in the system under test.

NEON LAMP

INCANDESCENT LAMP

RF VOLTMETER

THERMOCOUPLE
MILLIAMMETER

SECONDARY WINDING
WITH LAMP

SECONDARY WINDING
WITH METER

FIGURE 4.35. WAVEMETER RESONANCE INDICATORS.

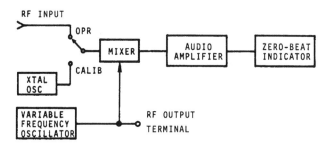

FIGURE 4.36. BLOCK DIAGRAM OF HETERODYNE FREQUENCY METER.

Wavemeters are sometimes classified as transmission types or as absorption types depending on the way they are connected to the frequency source being checked. A transmission-type wavemeter has connectors at both input and output ends so it can be put into the circuit under test. An absorption-type (or reaction-type) meter, on the other hand, is designed for loose coupling via the electromagnetic field of the frequency source.

The simplest LC absorption wavemeter for measuring radio frequencies consists of a coil and a variable capacitor in series. The capacitor shaft has a calibrated dial so the resonant frequency of the LC circuit can be read off as the capacitor is adjusted.

The simple wavemeter is used by holding it near the frequency source being measured. It is then tuned while watching for a sharp reaction by the source. The frequency is read from the calibrated dial. The reaction is caused by the absorption of a small amount of rf energy from the source by the wavemeter when the two are at the same frequency. In measuring the frequency of a small transmitter, one might look for a flicker in the power stage current as the wavemeter is tuned through resonance. Because of the loose coupling and the small power consumed by the wavemeter, the energy taken from high-power sources might go unnoticed; therefore, an indicator is built into the wavemeter itself.

A small lamp is sometimes used with the wavemeter for rough measurements of frequencies at high power levels. A neon lamp in parallel with the wavemeter's tuned circuit or an incandescent lamp in series with the tuned circuit will glow if enough power is absorbed from the frequency source. Better measurements can be made if the lamp is replaced by a meter. Figure 4.35 depicts indicator arrangements used with absorption wavemeters.

In operation, the wavemeter is tuned for maximum brilliance of the lamp or for peak deflection of the meter. A resonance indicator in series with the tuned circuit must have very low impedance so as not to degrade the Q (fractional linewidth), and hence the resolution, of the wavemeter. An indicator connected across the tuned circuit must have a high impedance for the same reason. The resonant elements themselves should exhibit the lowest losses possible, since the accuracy with which frequency can be measured by these means is mainly dependent on the Q of the tuned circuit.

4.4 HETERODYNE FREQUENCY METERS

The heterodyne frequency meter is basically a mixer, one input of which is driven by a reference signal from a self-contained, calibrated, tunable oscillator. It also includes an audio amplifier with means for operating a zero-beat indicator, such as headphones, an ac voltmeter, or an oscilloscope. Utilizing the familiar heterodyne principle, it compares the internal oscillator frequency with the unknown frequency of an external source.

The unknown signal is applied through the input terminals to the mixer stage, where it combines with the reference signal to produce a beat note. The internal oscillator is then carefully tuned for a zero beat. Its frequency is then read directly from the calibrated tuning dial. If the unknown frequency falls within the fundamental tuning range of the oscillator, its value will be the same as the oscillator frequency under the zero-beat conditions.

Because the nonlinear mixer gives rise to harmonics, it is also possible to detect beat notes between the fundamental of one signal and harmonics or subharmonics of the other. If the frequency of the external signal matches a known harmonic of the oscillator fundamental, the dial setting must be

57

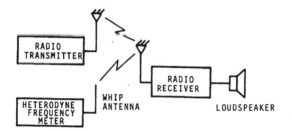

FIGURE 4.37. ARRANGEMENT FOR CHECKING TRANSMITTING FREQUENCY
WITH HETERODYNE FREQUENCY METER AND RECEIVER.

multiplied by the appropriate harmonic number N. On the other hand, if the unknown frequency equals a subharmonic, the dial reading must be divided by N. Thus, from a knowledge of harmonic relations, heterodyne measurements can be made of frequencies far above or below the fundamental tuning range of the frequency meter.

Figure 4.36 shows the block diagram of a basic heterodyne frequency meter. The internal variable-frequency oscillator must have good stability. This calls for close temperature regulation, complete shielding, and rugged construction. The oscillator should also be thoroughly buffered to reduce frequency-pulling tendencies when strong signals are applied from the external source, or when other factors would cause the oscillator load to change.

Some heterodyne meters are equipped with a whip antenna so the internal oscillator signal can be radiated over short distances. For checking the frequencies of transmitters or other high-power sources, a safe method is to tune in the transmitter with a radio receiver and then adjust the frequency meter until a zero beat is heard through the receiver's loudspeaker. The unknown frequency can be measured just as though it were coupled directly to the frequency meter.

The overall accuracy of measurement depends on the calibration accuracy of the tunable oscillator and the resolution attainable through the zero-beat process. At high frequencies, the error introduced by imperfect zero-beat detection can be ignored. Provided the temperature is kept nearly constant, the variable-frequency oscillator may retain its accuracy within 0.1% over long periods of time without recalibration.

Most heterodyne meters contain a stable quartz crystal oscillator for use in generating markers at fixed check points on the tuning dial. By checking the dial calibration at the point nearest the frequency to be measured, it is possible to achieve an accuracy approaching one part per million after the instrument has reached a stable operating temperature.

Heterodyne frequency meters combine wide frequency coverage with good accuracy in a portable instrument. They can be obtained for operating over fundamental frequency ranges from approximately 100 kHz to 200 MHz. Accuracy varies from about 0.01% at low frequencies to 0.00025% or better at very high frequencies. Harmonics up to 3 GHz may be generated with a precision of 1×10^{-6}.

Because they are tunable devices, the heterodyne frequency meters offer better noise immunity and signal sensitivity than wideband instruments such as oscilloscopes and counters. The heterodyne meter is one of the most useful of all frequency-measuring instruments for checking transmitter frequencies, for calibrating rf signal generators, and for other applications which do not require a high degree of signal purity.

4.5 DIRECT-READING ANALOG FREQUENCY METERS

Analog frequency meters operate by transforming the unknown frequency into a different physical quantity--such as voltage or current amplitude--which can be measured directly by conventional instruments. Thus, the output indicator might be an ordinary milliammeter whose deflection is somehow made proportional to the input frequency.

4.5.1 ELECTRONIC AUDIO FREQUENCY METER

A meter-type instrument that reads directly in audio frequency units and requires no manipulations of any kind is shown in figure

58

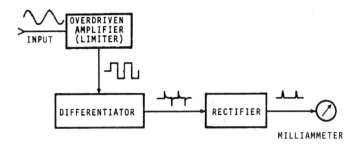

FIGURE 4.38. ELECTRONIC AUDIO FREQUENCY METER.

4.38. The input signal is shaped into a rec-tangular waveform, differentiated, rectified, and then applied to a milliammeter. The current spikes have the same frequency as the input signal; but since their peak amplitude is constant, the meter deflection is propor-tional only to the number of spikes that occur per unit of time. Deflection, therefore, is determined by the frequency of the input signal so the meter scale can be calibrated directly in hertz.

Operation of the instrument is no more complicated than the operation of an elec-tronic voltmeter. Besides its simplicity of operation, the instrument has an additional advantage in that frequency indications are independent of both waveform and amplitude of the input signal over very wide limits. Its overall accuracy is comparable to that of the meter movement, typically 1 to 5 percent.

4.5.2 RADIO FREQUENCY METER

The operating range of the audio frequen-cy meter may be extended to the VHF region or higher by adding a series of decade frequency scalers ahead of the instrument. Figure 4.39 shows a diagram of such a device which is

capable of measuring radio frequencies up to 50 MHz directly.

The audio frequency meter is preceded by four mixer stages, each excited by a signal derived from a standard 1 MHz crystal oscil-lator. The injection frequency for the first mixer consists of the 25th to the 49th har-monic of 1 MHz as determined by the tuning of harmonic selector A.

The difference between the input frequen-cy and the selected harmonic is then applied to a second mixer, where it is heterodyned against a subharmonic of 1 MHz as determined by the tuning of harmonic selector B. The process continues until the input frequency is eventually reduced to a value that falls within range of the direct-reading audio meter.

The instrument is operated by manually tuning each harmonic selector and mixer in succession for peak deflection on the associ-ated tuning meter. Each harmonic selector dial is calibrated so that the input frequency can be read directly from the dial settings in conjunction with the reading of the audio meter. For example, if optimum tuning occurs when dial A is set at 32, dial B is at 7, dial C is at 5, dial D is at 9, and the audio meter

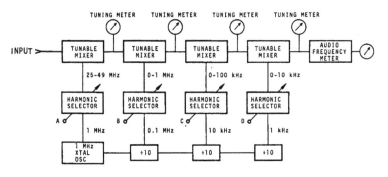

FIGURE 4.39. DIRECT-READING RADIO FREQUENCY METER.

59

is reading 170, then the input frequency is 32.759170 MHz. Typical accuracy of the instrument with an analog audio meter is 5 x 10⁻⁷ at VHF. If the audio meter is replaced by an electronic digital counter, the accuracy may be improved to 2×10^{-8} or better.

4.6 FREQUENCY COMPARATORS

Frequency comparators are versatile devices for making frequency measurements and comparisons. These instruments combine phase error multipliers, phase detectors, integrated circuits and frequency multipliers into useful and accurate laboratory aids. The block diagram of this instrument is shown in figure 4.40. The heart of the frequency comparator is a variable-multiplication phase error multiplier of special design.

The phase error multiplier, like all electronic devices, is subject to noise. Through each stage of error multiplication, the noise at the input is amplified. Eventually the noise out of the phase error multiplier would be stronger than the signal, so to maintain a useful signal-to-noise ratio the multiplication is usually limited to 10,000.

Probably the most important part of this instrument is the frequency discriminator which is based on a tuned-circuit. The output is a dc voltage proportional to the difference between the frequency of the discriminator input signal and the resonant frequency of the tuned circuit. The Q of the tuned circuit determines the sensitivity of the discriminator.

As an example of the Q required, consider a discriminator operating at 1 MHz with an error multiplication of 10,000. If the comparator is to respond to a frequency difference of 1 part in 10^{12}, the discriminator will have to detect a frequency difference of $\Delta f = 10^{-12} \times 10^4 \times 10^6$ Hz = 0.01 Hz. If 0.01 Hz corresponds to 1% of full-scale meter deflection, the bandwidth of the discriminator must be 100 x 0.01 Hz = 1 Hz. This requires a Q of

10^6. One commercial comparator solves this problem by using a quartz crystal as the tuned circuit for the discriminator.

The comparator usually contains phase detectors that also are connected to the output of the phase error multiplier chain. The phase detectors have various integrating times, or bandwidths, that allow one to optimize the accuracy of the particular measurement being made. As shown in figure 4.40, the comparator can be used with an external chart recorder, an electronic digital counter, or an oscilloscope.

4.7 AUXILIARY EQUIPMENT

Some specialized types of equipment, though not complete systems for the measurement of frequency or time, are nevertheless very useful as auxiliary devices to increase the versatility, accuracy, and range of other measuring instruments. These auxiliary devices include frequency synthesizers, phase error multipliers, phase detectors, frequency multipliers, frequency dividers, signal averagers, and phase-locking receivers.

4.7.1 FREQUENCY SYNTHESIZERS

Not all frequency comparisons and measurements are performed at even frequencies like 100 kHz, 1 MHz, 5 MHz, etc. Many electronic applications involve odd frequencies. An example is the television color subcarrier frequency of 63/88 x 5 MHz. VHF and UHF communications channel frequencies are another example.

There are two general types of synthesizers, each with points in its favor. The direct synthesizer uses internal frequency multipliers, adders, subtracters, and dividers in combination to produce an output frequency. The indirect method of synthesis uses a phase-lock oscillator to provide the actual output signal. The spectral purity of the latter

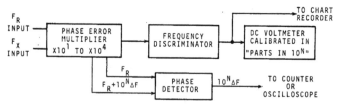

FIGURE 4.40. BASIC FREQUENCY COMPARATOR.

ethod is more a function of the phase-lock
oop than of the spectrum of the locking
ignal. Both types of synthesizers are widely
sed in modern electronic equipment for the
eneration of frequencies.

The synthesizer has two specifications
hich determine its usefulness in any appli-
ation. One is <u>range</u> and the other is <u>reso-
ution</u>. If the instrument has a range of 10
Hz and a resolution of 0.1 Hz, any frequency
p to 10 MHz may be generated in steps of 0.1
z. The dials or buttons on the front of the
nstrument display the output frequency.
hese controls interconnect the various
ultipliers, mixers, etc. that are necessary
o produce the output frequency.

As mentioned previously, the measurement
nd comparison of odd frequencies is simpli-
ied by the use of a frequency synthesizer.
s an example, assume that one wishes to
etermine the frequency of a crystal oscil-
ator. The output frequency may be measured
ith an electronic counter if one is avail-
ble. Barring the availability of a counter,
Lissajous pattern may be displayed on an
scilloscope with some reference frequency;
ut the pattern on the oscilloscope might be
oo complex for an accurate interpretation.

On the other hand, if a frequency syn-
hesizer is available, a low-ratio Lissajous
attern may be obtained. Once the pattern is
tabilized, the oscillator frequency may be
ead off the synthesizer dials. If the pat-
ern is other than 1:1, the oscillator fre-
uency may be computed by using the dial value
d the pattern ratio. Similarly, phase or
requency stability of odd-frequency oscil-
tors may be determined using a synthesizer
conjunction with phase detectors, oscil-
scopes and chart recorders.

4.7.2 PHASE ERROR MULTIPLIERS

Oscilloscopes, phase detectors, chart
recorders, and similar devices used to make
frequency comparisons may be limited by their
frequency response. This can be a serious
handicap when frequency multiplication is used
to speed up calibrations. A phase error
multiplier, however, multiplies the frequency
<u>difference</u> between two signals without chang-
ing the frequencies themselves.

A phase error multiplier is shown in
figure 4.41. It consists of one or more
decade frequency multipliers in conjunction
with one or more mixers whose outputs are the
difference frequency between their input
signals.

The reference frequency f_r is multiplied
by nine to produce $9f_r$. The unknown frequency
f_x is multiplied by ten to produce $10f_x$. But
f_x may be redefined as $f_x = f_r \pm \Delta f$, where Δf
is the difference frequency. Then $10f_x = 10f_r$
$\pm 10\Delta f$, and the mixer output will be $10f_r \pm$
$10\Delta f - 9f_r = f_r \pm 10\Delta f$.

The difference frequency is now 10 times
what it was initially. This process can be
repeated, as shown in figure 4.41, until the
noise gets too big. Four such decades seem to
be about the limit. This is equivalent to
multiplying two 1-MHz signals to 10 GHz. The
output of the phase error multiplier is now
treated the same way as the input would have
been, but the greatly amplified frequency
difference simplifies and quickens the measurement process. An application of the phase
error multiplier is discussed in the sections
on Lissajous patterns (4.2.3A) and frequency
comparators (4.6).

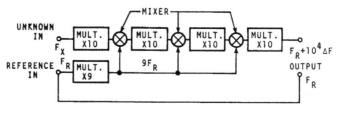

$$\Delta F = |F_R - F_X|, \quad F_X = F_R \pm \Delta F$$

FIGURE 4.41. PHASE ERROR MULTIPLIER.

4.7.3 PHASE DETECTORS

Why use a phase detector? If two oscillators differed by 1 Hz and we compared their frequencies, we'd see a 1 Hz signal. But what if they only differed by 0.001 Hz? A chart record of their difference frequency or "beat note" would take 1000 seconds to trace a cycle. We soon find ourselves dealing with fractions of a cycle. A cycle has 360 electrical degrees. If we measure the phase of the two signals, then we can calibrate our chart or oscilloscope in terms of phase. This is simply a way of looking at fractional hertz--parts of a cycle.

For example: Consider two 1 MHz oscillators that differ in frequency by 1 part in 10^8. This means that it will take 100 seconds for a cycle of their beat note to occur. On a chart, we would see 1 cycle (equals 1 microsecond at 1 MHz) in 100 seconds. In 10 seconds, only one-tenth of a microsecond change will have occurred. These small changes of phase can be used to calibrate frequency differences. Of course, if the signals we are comparing differ by large amounts, the phase changes very rapidly.

There are two main types of phase detectors--the linear type and the nonlinear type. They are discussed separately in the following paragraphs. The linear phase detector gives a linear output as the relative phase of the input signals change. They are often used with phase error multipliers and strip chart recorders. In a previous section, it was noted that linear phase detectors often cannot accept very high frequencies. This is a good reason for combining a phase detector with a phase error multiplier.

A block diagram of a typical linear phase detector is shown in figure 4.42a. Examination shows that the heart of this type of phase detector is a flip-flop. The action, or duty cycle, of the flip-flop determines the output voltage of the detector. An upper frequency limit for the detector occurs because of the flip-flop. If the shaped signals from the two inputs are less than 0.1 microsecond apart, for instance, the flip-flop might not react at all. So the dc voltage to the chart recorder will not be a true time record of the phase difference. In fact, what happens is that the sawtooth output shown in figure 4.42b begins to shrink. The waveform shifts up from zero and down from full-scale. This shrunken response could be calibrated, but it is much more convenient to use the phase detector at a frequency which it can

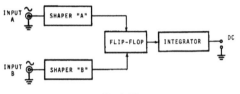

a. Block Diagram.

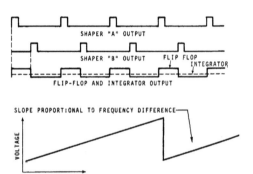

b. Waveforms.

FIGURE 4.42. LINEAR PHASE DETECTOR.

handle. This is where a phase error multiplier proves helpful. With the phase error multiplier, one is able to have very sensitive full-scale values of phase shift, say 0.01 microsecond, with inputs in the 100-kHz range.

A nonlinear phase detector is essentially a phase discriminator like that used in FM receivers. The circuit diagram of a typical nonlinear detector is presented in figure 4.43. These detectors are available with cable connectors or in a printed circuit board package. They are very small.

Because the response characteristic is nonlinear, this phase detector is not used extensively with chart recorders. It has, however, one important characteristic--its output voltage can reverse polarity. In other

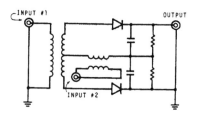

FIGURE 4.43. SCHEMATIC DIAGRAM OF NONLINEAR PHASE DETECTOR.

words, the output is a dc voltage proportional to the sine of the phase difference between the input signals, and the polarity of the output voltage depends on the lead or lag condition of the reference signal with respect to the unknown signal phase.

If the input signals are different in frequency, the output voltage is periodic like a sine wave and there is no dc component. This ac output can be measured with a frequency counter, or it can be applied to an oscilloscope to form a Lissajous pattern.

There are commercial meters available to make phase measurements. Several of these can compare signals at very high frequencies directly without the need to use phase error multipliers, etc. By using sampling techniques, comparisons up to 1000 MHz can be made. Read the equipment manuals carefully to note the linearity and accuracy of any instrument so that the results can be correctly understood.

4.7.4 FREQUENCY DIVIDERS

Frequency dividers are special forms of frequency synthesizers. They may be grouped into two general types: analog dividers that use regenerative feedback, and digital dividers that use binary arithmetic together with logic functions provided by bistable circuits.

Analog dividers lend themselves to rf applications in conjunction with receivers and transmitters. Digital types are found in electronic counters, computers and slewable dividers.

A. Analog or Regenerative Dividers

Figure 4.44 shows the block diagram of an analog or regenerative divider. Frequency f is supplied to the input of the divider. After amplification, this signal becomes one of two inputs to a mixer stage. The mixer output at f/10 is multiplied to 9f/10 and then fed back to the other input of the mixer.

The gain of the various stages is very critical. The regenerative divider is actually an oscillatory circuit though not self-oscillating. If the loop gain around the feedback circuit is too high, the unit will continue to divide even with no input. The output frequency in this condition is determined by tuning of the various circuits. If the circuit gain is too low, the unit will not function at all. The proper condition, as one would suppose, is where the unit divides under conditions of input signal but ceases operation when that input signal is removed.

Some commercial units have their gain set so that the dividers are not self-starting. This feature is designed to eliminate the problems of division with no input signal. Also, since the unit isn't self-starting, the output phase should not shift due to momentary loss of input or dc power. Once the divider stops, it remains stopped until a start button is pressed.

B. Digital Dividers

Simple bistable circuits such as flip-flops may be combined to perform digital division. Any flip-flop, whether assembled from discrete components or integrated circuits, performs the same basic function. The output assumes two possible states, first one

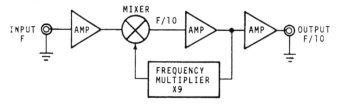

FIGURE 4.44. ANALOG OR REGENERATIVE TYPE OF DECADE FREQUENCY DIVIDER.

63

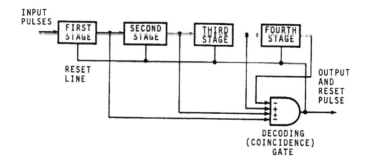

FIGURE 4.45. BLOCK DIAGRAM OF A FOUR-STAGE FLIP-FLOP DIVIDER.

and then the other, when the circuit is triggered by two successive input pulses. Thus,
we have division by two.

If the output of one flip-flop is connected to the input of another, the output of
the second flip-flop will be a division by
four. A third flip-flop would allow division
by eight, and so on. Each succeeding flip-
flop provides an additional division by two.
The divisors are 2, 4, 8, 16, 32, etc.

A decade (÷10) divider is formed by using
extra circuitry. One such divider uses internal feedback to supply six extra pulses
to a four-stage flip-flop chain, thereby
forcing the circuit into supplying an output
pulse after only ten input pulses rather than
the normal sixteen.

New digital logic circuits are being
produced so rapidly and at such low cost that
we cannot list all the possibilities here.

Every manufacturer has literature available
that gives applications for their circuits.
You can assemble dividers, synthesizers, and
even digital crystal oscillators to give you
just about any frequency and waveform you
desire. There are circuit "chips" that will
generate triangular waves and pulse bursts.
Complete phase-lock loops are available in
small packages. There are low-power circuits
and high voltage dividers and drivers for a
variety of indicators and displays.

4.7.5 ADJUSTABLE RATE DIVIDERS

When using an oscilloscope to observe
periodic waveforms, such as pulses or short
bursts of sine waves extracted from radio
receivers, it is often advantageous to trigger
the oscilloscope from a separate source. The
problem arises as to how to position the
pattern for best viewing or measurement. An

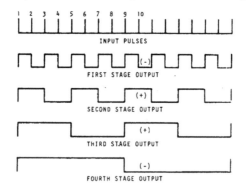

FIGURE 4.46. FOUR-STAGE FLIP-FLOP DIVIDER WAVEFORMS..

64

adjustable digital divider can solve this problem.

Assume that we have a digital divider which accepts a 1-MHz input signal, counts 10^6 cycles of this input, and then delivers a narrow output pulse. If this process is repeated continuously, the output pulses will occur once every second. If we use the 1-pps output to trigger an oscilloscope, we may need to shift this pulse around in time to see a desired waveform.

If during the counting cycle, we were to add an extra pulse to the input pulse train, the dividers would count this additional pulse along with the normal pulses. The divider chain would count along as usual until the number 1,000,000 is reached. At that instant, the unit would send out its normal seconds pulse. But since we added one extra pulse to the input train, the output pulse would occur 1 microsecond earlier than it would have otherwise. If we add a pulse to the input train every second, the output pulse would advance in time by 1 microsecond per second. If we subtract one pulse from the input train each second, the output pulse would be retarded 1 microsecond per second. By adding or subtracting pulses at different rates, we can make the output pulses move rapidly or slowly forward or backward. This is how the adjustable divider works. It can be built up easily with modern digital circuitry.

4.7.6 SIGNAL AVERAGERS

If the same point on the seconds pulse is examined each time the pulse occurs, an average voltage for that point will emerge. This is because the signal amplitude at the point is constant, and ultimately the random noise voltage at that point will average out to zero. The length of time required for the average signal value to appear will depend on the amount and nature of the noise.

The signal averager examines many points on a pulse. The instantaneous value of voltage at each point is stored in a memory circuit comprised of a bank of high-quality capacitors. Each time the pulse occurs, the same points are examined and stored in the same memory elements. After a sufficient length of time, each capacitor will have stored the average value of voltage from its respective point on the pulse.

Each of the memory capacitors is in series with an electronic switch. The switches are closed and opened by a control circuit, which operates each switch in succession so that only one is closed at any given time. Resistors between the input and output amplifiers allow the time constant of the memory elements to be optimized with respect to the length of the pulse.

As each switch is closed, its associated capacitor charges to the instantaneous voltage present at the point on the waveform being examined. That voltage is simultaneously displayed on an oscilloscope. The waveform of the displayed pulse is composed then of many discrete voltage levels read from many storage capacitors.

As one can imagine, the entire process must be carefully synchronized. The switch controller is triggered simultaneously with the oscilloscope trigger, and the switching rate is adjusted nominally to allow a display of all capacitor voltages during one sweep of the oscilloscope beam.

The signal averaging technique allows one to see weak signals in noise without loss of signal bandwidth. Commercial instruments are available using the analog method described above and also digital techniques. After the signal has been recovered from the noise,

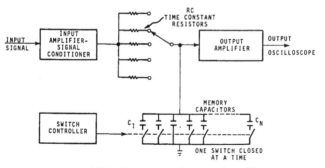

FIGURE 4.47. SIMPLIFIED DIAGRAM OF A SIGNAL AVERAGER.

65

measurements can be made to determine the time of arrival, etc. A local clock "tick" can be used to trigger the start of the averaging process.

4.8 PHASE LOCK TECHNIQUES

There are many books and articles on this subject. As mentioned, even integrated circuits are available to phase lock signals. Here, we only want to acquaint the reader with the concept of the phase-locked loop and what it will do for our time and frequency measurement problems. If further information is desired, the reader should seek out texts and application notes on the subject.

The phase-locked loop is made by connecting a phase comparator, integrator, and voltage-controlled oscillator (VCO). Any one of these blocks can be implemented in many ways. Functionally, a crystal oscillator that reacts to an input voltage by changing its frequency is the same as a dc motor that speeds up or slows down as its input voltage is changed.

The net effect of a phase-lock loop can be one of several things. It can change the input frequency. For example, if the input is 1 MHz, we could put a 100 kHz oscillator in for a VCO and a multiplier in the circuit and have a divider. Or, we could use a 10 MHz VCO and a divider in the circuit and have a multiplier. We could also change the power level. That is, a very small input signal could be made to control a powerful output stage. This would give us a power amplifier.

Typical applications for a phase-locked loop involve the idea of filtering the input signal. Usually a low-pass filter does this for us. The VCO is connected after the filter. This prevents high frequency signals (where the noise usually is) from affecting the VCO. So, if the input is a 1 MHz signal with noise, we can remove the noise and keep the 1 MHz signal. It's not really that easy or simple, but just an example of what is happening.

Since very narrow bandwidths can be achieved, we could separate one signal from another even though they are very close together. If you read articles written in the field of time and frequency technology, mention is often made of using phase-locked loops to remove noise from signals after they have arrived over telephone lines, etc.

An important application of phase-lock techniques is in the VLF tracking receiver. The application of the loop is to measure the phase of a local oscillator with respect to a received signal at VLF frequencies. Here the phase-locked loop filters provide an analog phase output and even synthesizes the necessary frequencies so the receiver can be tuned throughout the VLF band.

The block diagram of a simple phase-tracking receiver is shown in figure 4.48. Basically, this receiver adjusts the phase of a local frequency standard to agree with that of a received radio signal. The phase adjustment is performed by a phase shifter, which also accepts the output signal of the local standard. This shifted signal becomes the input signal for an internal frequency synthesizer whose frequency is set equal to the received frequency. Comparison between these two signals is then made by a phase detector which in turn controls the action of the phase shifter in such a way as to maintain the phase difference, ε, at a nearly zero value.

A time constant in the servo loop limits and controls the rate of phase change that can be accomplished by the shifter. This time constant permits the receiver to have a very narrow bandwidth. Commercial receivers feature adjustable time constants which allow the effective bandwidth to be reduced to as low as

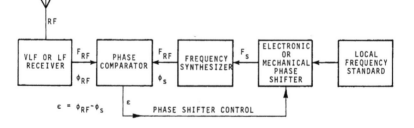

FIGURE 4.48. SIMPLE PHASE-TRACKING RECEIVER.

0.001 Hz. With bandwidths this narrow, the receiver can follow a signal that is much weaker than the neighboring noise levels.

Another way to view the action of a phase-tracking receiver is to consider it a signal-averaging process. The long time constants (100 seconds or more) characteristic of these receivers may be thought of as averaging times. The phaselocked output frequency of the receiver is a faithful representation of the received frequency and phase. Only the noise has been removed.

4.9 SUMMARY

There are many types of equipment available for use in making time and frequency calibrations. Perhaps the most important are the electronic counter and the oscilloscope. There are also many techniques that can be used, depending on the application. We again encourage users to discuss their needs with manufacturers of time and frequency equipment and to read the available application notes and other literature.

Now that we have discussed the use of time and frequency in the lab, we will turn our attention to some of the services that provide accurate signals for calibration. These services range from high and low frequency radio broadcasts to newer services that use network television as a transfer standard to satellites.

Obviously, you will not want to use all of the services. The magnitude of your measurement problems, your accuracy requirements, the manpower available, and your budget are all contributing factors in choosing a service that best meets your needs. Hopefully, the following chapters will provide enough information to allow you to make a wise decision.

CHAPTER 5. THE USE OF HIGH-FREQUENCY RADIO BROADCASTS
FOR TIME AND FREQUENCY CALIBRATIONS

High frequency (HF) shortwave radio broadcasts are one of the most popular sources of time and frequency information. HF signals from stations such as WWV (Ft. Collins, Colorado), WWVH (Kauai, Hawaii), and CHU (Ottawa, Canada) are readily available and provide essentially worldwide coverage. In addition, the signals can be picked up with relatively inexpensive receiving equipment.

Although HF broadcasts are extremely popular, there are drawbacks to using them. To date, no practical method has been found to automate the use of HF signals. Also, the equipment requires human operators.

HF signals rely primarily on reflection from the sky (ionosphere) to arrive at a distant point. Changes in the density and height of the ionosphere (or reflecting region) change the distance a time signal must travel. This produces a characteristic of high frequency or "short" waves called fading. The signals that leave the transmitter take a number of paths to reach the receiver. As these signals arrive at a distant point, they have shifted in phase by different amounts. Sometimes they recombine to produce a stronger signal, but at other times, they almost cancel each other and no signal is available.

It is this recombining-canceling pattern that produces the characteristic fading at HF. This also changes the apparent time of a received pulse or "tick." The arrival time of a timing pulse may vary from day to day, even for measurements made at the same time of day. Accuracies approaching 1 millisecond can typically be achieved by taking measurements each day and averaging over several days. This means that a user must have a clock and the necessary equipment to keep good time and measure small time differences. It also means he must be dedicated and willing to expend the effort to do the job.

Good signal reception is essential for accurate frequency or time measurement using HF broadcasts. Signal strength and geographical location will determine the necessary receiver and antenna requirements. For example, a directional antenna may be necessary so that it can be oriented to favor the transmission mode which consistently provides the strongest received signal. Accuracy of measurements using HF signals is typically 10 milliseconds for time and 1×10^{-6} for frequency.

5.1 BROADCAST FORMATS

As previously mentioned, there are many HF radio stations worldwide that can be used for time and frequency measurements. Some of them are described in this chapter.

5.1.1 WWV/WWVH

Standard time and frequency stations WWV and WWVH broadcast on carrier frequencies of 2.5, 5, 10, and 15 MHz. WWV also broadcasts on 20 MHz. The stations get all of their signals from a cesium beam frequency source. They each use three "atomic clocks" to provide the time of day, audio tones, and carrier frequencies. The rates or frequencies of the cesium oscillators at the stations are controlled to be within one part in 10^{12} (1000 billion) of the NBS frequency standard located in Boulder, Colorado. Time at the stations is held to within a few microseconds of the NBS atomic time scale, UTC(NBS).

These cesium standards drive time code generators through various dividers, multipliers, and distribution amplifiers. The time code generators generate audio tones and time ticks. Frequency multipliers provide the radio carrier frequencies which are then amplitude modulated.

The seconds pulses or "ticks" transmitted by WWV and WWVH are obtained from the same frequency source that controls the carrier frequencies. They are produced by a double sideband, 100 percent modulated, signal on each RF carrier. The first pulse of every hour is an 800-millisecond pulse of 1500 Hz. The first pulse of every minute is an 800-millisecond pulse of 1000 Hz at WWV and 1200 Hz at WWVH. The remaining seconds pulses are brief audio bursts (5-millisecond pulses of 1000 Hz at WWV and 1200 Hz at WWVH) that

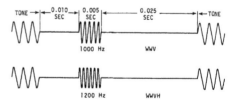

FIGURE 5.1. FORMAT OF WWV AND WWVH SECONDS PULSES.

69

sound like the ticking of a clock. All pulses occur at the beginning of each second. The 29th and 59th seconds pulses are omitted.

Each tick is preceded by 10 milliseconds of silence and followed by 25 milliseconds of silence to avoid interference from other tones and to make it easier to hear the tick. The total 40-millisecond protected zone around each seconds pulse is illustrated in figure 5.1. This means that the voice announcements are also interrupted for 40 milliseconds each second. This causes only a small audio distortion. The ticks have priority and must be received clearly.

Transmitting equipment at each station consists of high-power linear amplifiers which are connected to separate antennas. The several WWVH antennas are directional arrays except for the 2.5 MHz antenna. All the antennas at WWV are omnidirectional. The directional property of the WWVH antennas is intended to minimize "interference" between the two stations and to improve coverage in the Western Pacific.

The complete broadcast format for WWV and WWVH, showing exactly what is broadcast during each minute of the hour, is shown on page 215 of Chapter 12. See Table 5.1 for other details.

5.1.2 CHU

Canada has many frequency and time services that are quite similar to those of the U. S. The Canadian HF broadcast station near Ottawa has the call letters CHU. Its signals can be heard over much of the U. S. and are a valuable alternative to the WWV signals. It has the same propagation characteristics and the same receiving techniques are used.

The CHU signals differ from those of the U.S. stations in two ways. They are not in

the standard frequency bands (see below) and they use a different format. The format is principally voice and ticks. At CHU, dual cesium frequency sources are used to generate the carriers and the seconds pulses. Two systems are used to complement each other in cases of maintenance or failure. The output from the cesium oscillator is fed to a solid state frequency synthesizer which produces the 3330, 7335, and 14670 kHz carrier signals for the transmitters.

The three excitation frequencies are fed into a transmitter room and terminated in a panel where each frequency is coupled into two transmitters (main and standby) operating on that frequency. A 100 kHz signal is fed into one of two digital clocks where it is divided into seconds pulses of 1000 Hz tone. The clock gates out the 51st to 59th second pulses of each minute to permit the voice announcements to be inserted, and also gates out the 29th pulse to identify the half-minute point.

The same 100 kHz signal is fed into a "talking clock" where it is divided down to 50 Hz and amplified to a power level that will operate a synchronous motor. The motor drives a magnetic drum on which the voice announcements are recorded. Two talking clocks are used. One announces the station identification and time in English, and repeats the time in French. The other announces the station identification and time in French, and repeats the time in English. The format is shown in figure 5.2.

In 1975, CHU converted its operation to single sideband. It now broadcasts upper sideband with full carrier. This is called the 3A3H mode. Users can still get the carrier as a frequency standard and an ordinary AM radio will allow reception of the audio signals.

A time code is now being tested on CHU. It appears in the 31st to the 39th seconds pulses of each minute. The modulation is the

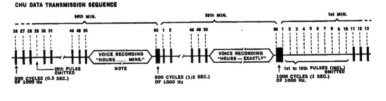

FIGURE 5.2. CHU BROADCAST FORMAT.

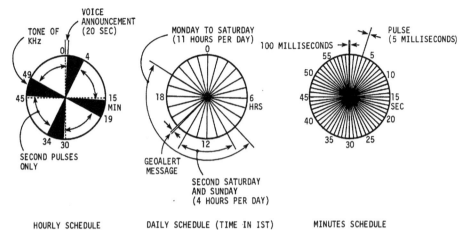

FIGURE 5.3. TRANSMISSION SCHEDULE OF ATA OVER A DAY, HOUR, AND MINUTE.

commercial 300 baud FSK (frequency shift keying) at frequencies of 2025 and 2225 Hz. The code is a modified ASCII in which each 11-bit character contains two BCD digits, with one start bit and two stop bits. The first digit is the number 6 for identification, which should be verified in the receiving clock, and the remaining 9 digits give the day, hour, minute and second of UTC. The entire message is then repeated in order that the receiving clock can check for identical messages before updating. The code ends, and update occurs at 0.500s. This 0.500s must be added in, along with the time zone hour, to give the correct time. The same code is also available by telephone. Details can be obtained by writing to the Time and Frequency Section, National Research Council, Ottawa, Ontario, Canada. Other details of CHU's service are shown in Table 5.1, page 77.

If you are outside the United States or Canada, the following HF stations are available for time and frequency measurements.

5.1.3 ATA

ATA, which broadcasts on 5, 10, and 15 MHz, is located in Greater Kailash, Delhi, India. It is operated by the National Physical Laboratory in New Delhi. The transmission schedule is shown in figure 5.3.

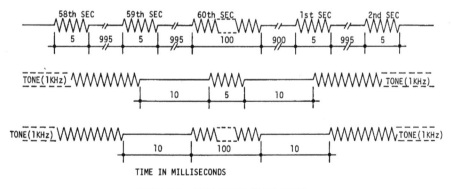

TIME IN MILLISECONDS

FIGURE 5.4. TIME FORMAT OF ATA SIGNAL.

71

ATA gives time ticks for every second, minute, and quarter-hour (i.e., 0th, 15th, 30th, and 45th minute of every hour). The second pulses consist of a group of 5 cycles of 1 kHz of 5 milliseconds duration. The minute pulses, started at the 0th second, are a 1 kHz signal of 100 milliseconds duration. At the beginning of every quarter-hour, a 1 kHz tone starts; it lasts for four minutes (i.e., from the 0th to 4th, 15th to 19th, 30th to 34th, and 45th to 49th minute).

The second and minute pulses are maintained during the tone period by interrupting the tone for 25 and 120 milliseconds respectively and introducing second and minute pulses. The second and minute pulses are preceded and succeeded by 10 millisecond intervals as shown in figure 5.4.

A voice announcement, comprising the station call sign and Indian Standard Time is made about 20 seconds prior to the beginning of each quarter-hour. Other details of ATA are shown in Table 5.1.

5.1.4 IAM

The Istituto Superiore delle Poste e delle Telecomunicazioni (ISPT) in Rome, Italy, operates HF radio station IAM on 5 MHz. IAM broadcasts time signals and standard frequencies.

The transmission is effected for 6 days a week (week-days only) in accordance with the diagram in figure 5.5.

The second pulse is made up of 5 cycles of a 1000 Hz standard frequency. Minute pulses consist of 20 cycles.

The UT1 correction is given by double pulses. Additional details are shown on Table 5.1.

5.1.5 IBF

IBF is operated by the Istitute Elettrotecnico Nazionale Galileo Ferraris in Torino, Italy. It transmits on 5 MHz.

Seconds pulses consist of 5 cycles of 1 kHz modulation. Seven pulses mark the minute. Voice announcements (given in Italian, French, and English) are given at the beginning and end of each emission. Time announcements are given by Morse Code every 10 minutes beginning at 0 hours 0 minutes.

The UT1 correction is given by double pulses. Additional details are shown on Table 5.1.

5.1.6 JJY

The Radio Research Laboratories in Tokyo, Japan, operate radio station JJY. Like WWV and WWVH, JJY broadcasts Coordinated Universal Time on 2.5, 5, 10, and 15 MHz. The types of information broadcast are also similar to WWV/WWVH.

JJY transmissions are continuous except for interruptions between minutes 25 and 34. The call letters are announced six times an hour at 9, 19, 29, 39, 49, and 59 minutes after each hour. The 1 Hz modulation pulses are of 5 milliseconds duration (8 cycles of 1600 Hz). The first pulse in each minute is preceded by a 600-Hz tone 655 milliseconds long. The format is modulated by 1000 Hz for intervals of 0 to 5, 10 to 15, 20 to 25, 30 to 35, 40 to 45, and 50 to 55 minutes, excluding the periods of 40 milliseconds before and after each second pulse. The hourly modulation schedule, waveform of second pulses, and identification of minute signal by preceding marker are shown in figure 5.6.

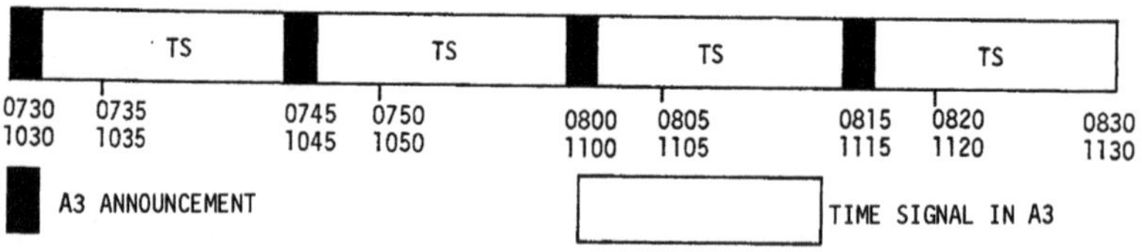

		TS		TS		TS		TS	

0730   0735   0745   0750   0800   0805   0815   0820   0830
1030   1035   1045   1050   1100   1105   1115   1120   1130

■ A3 ANNOUNCEMENT   ☐ TIME SIGNAL IN A3

TIME IS GIVEN IN SLOW-SPEED TELEGRAPHY AT 0735, 0750, 0805, 0820, 1035, 1050, 1105, AND 1120 HRS UTC.

DURING SUMMER TIME, THE EMISSIONS ARE ADVANCED BY 1 HOUR.

FIGURE 5.5. TRANSMISSION SCHEDULE OF IAM.

Elettro-
n Torino,

cles of 1
e minute.
, French,
nning and
ments are
beginning

by double
a on Table

in Tokyo,
Like WWV
Universal
e types of
imilar to

ous except
25 and 34.
x times an
59 minutes
:ion pulses
cycles of
a minute is
illiseconds
1000 Hz for
o 25, 30 to
, excluding
before and
·ly modula-
pulses, and
, preceding

0830
1130

105,

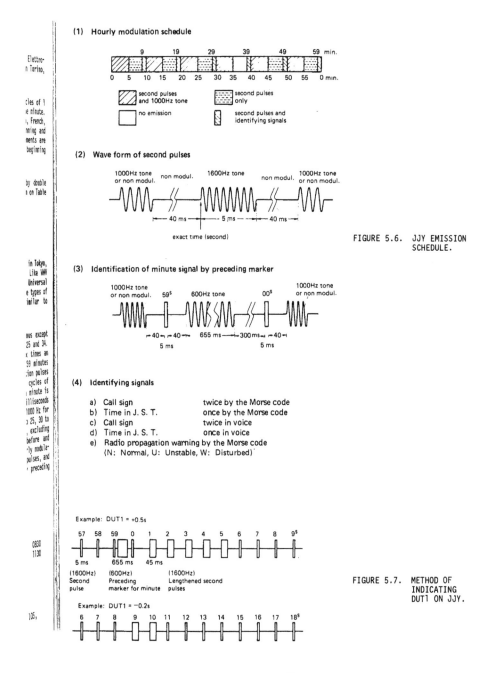

(1) Hourly modulation schedule

9   19   29   39   49   59 min.

0 5 10 15 20 25 30 35 40 45 50 55 0 min.

second pulses and 1000Hz tone
no emission
second pulses only
second pulses and identifying signals

(2) Wave form of second pulses

1000Hz tone or non modul.   non modul.   1600Hz tone   non modul.   1000Hz tone or non modul.

—40 ms—|— 5 ms —|—40 ms—

exact time (second)

(3) Identification of minute signal by preceding marker

1000Hz tone or non modul.   59^{s} 600Hz tone   00^{s} 1000Hz tone or non modul.

—40—|—40—   655 ms—|—300 ms—|—40—
5 ms   5 ms

(4) Identifying signals

a) Call sign   twice by the Morse code
b) Time in J. S. T.   once by the Morse code
c) Call sign   twice in voice
d) Time in J. S. T.   once in voice
e) Radio propagation warning by the Morse code
   (N: Normal, U: Unstable, W: Disturbed)

Example: DUT1 = +0.5s

57 58 59 0 1 2 3 4 5 6 7 8 9^{s}

5 ms   655 ms   45 ms
(1600Hz) (600Hz)   (1600Hz)
Second   Preceding   Lengthened second
pulse   marker for minute   pulses

Example: DUT1 = −0.2s

6 7 8 9 10 11 12 13 14 15 16 17 18^{s}

FIGURE 5.6.   JJY EMISSION SCHEDULE.

FIGURE 5.7.   METHOD OF INDICATING DUT1 ON JJY.

DUT1 is given once each minute to the nearest 0.1 second. The magnitude and sign are indicated by the number and position of the lengthened second pulses of 45 ms duration of 1600 Hz. Identifying signals and the method of indicating DUT1 are shown in figures 5.7. For additional details, see Table 5.1.

5.1.7 OMA

Station OMA in Liblice, Czechoslovakia, operates on 2.5 MHz. The station transmits seconds, minutes, and five-minutes marks formed by interrupting the 1-kHz subcarrier for 5 milliseconds, 100 milliseconds, and 500 milliseconds, respectively. The modulation schedule, as follows, is repeated each hour:

Minute 00-01, 15-16, 30-31, 45-46:

10 times OMA in Morse Code, 800 Hz tone.

Minute 01-05, 16-20, 31-35, 46-50:

1 kHz standard frequency transmitted from 6 hours to 18 hours UT only; replaced by seconds pulses the rest of the day.

Minute 05-15, 25-30, 35-40, 50-60:

Seconds pulses 5 milliseconds long, each 60th pulse is 100 milliseconds, each 300th pulse is 500 milliseconds. From 18 hours to 6 hours UT, the 5 millisecond pulses replace the 1 kHz tone.

Minute 20-25:

Carrier frequency transmission without modulation.

Minute 40-45:

Transmission is interrupted.

At the end of the 15th, 30th, 45th, and 60th minute of the hour, a short time signal is transmitted consisting of six second pulses 100 milliseconds long. See Table 5.1 for further details.

5.1.8 VNG

The Australian Telecommunications Commission provides HF standard frequency and time signals broadcasts from radio station VNG at Lyndhurst, Victoria. This service transmits precise time signals and radio frequencies on a schedule designed to provide an Australia-wide coverage. The transmitting schedule is shown below.

SCHEDULE OF FREQUENCIES

TIMES OF EMISSION			FREQUENCY
UTC	AEST	AEDT	MHz
0945-2130	1945-0730	2045-0830	4.5
2245-2230	0845-0830	0945-0930	7.5
2145-0930	0745-1930	0845-2030	12.0

UTC: Universal Coordinated Time
AEST: Australian Eastern Standard Time
AEDT: Australian Eastern Daylight Time

A. Time Signals

Seconds markers are transmitted by double-sideband amplitude modulation of the carrier and consist of various length bursts of 1000 Hz tone. The start of a tone burst, which commences at a zero-crossing of the sine wave (increasing carrier power), marks the start of a seconds interval.

B. Time Code

Seconds markers are normally 50 milliseconds of 1000 Hz. Seconds markers 55 to 58 are 5 milliseconds of 1000 Hz. Seconds marker 59 is omitted.

Minute marker (seconds marker 60) is 500 milliseconds of 1000 Hz. During the 5th, 10th, 15th, etc. minutes, seconds markers 50 to 58 are 5 milliseconds of 1000 Hz.

During the 15th, 30th, 45th, and 60th minutes, a station identification announcement is given without interruption to the time signals. During the first 15 seconds of each minute, a group of emphasized seconds markers (emphasized by 50 milliseconds of 900 Hz tone immediately following the normal seconds markers) gives the UT1 correction. The time code format is shown in figure 5.8.

C. Announcements

The announcement gives station identification (call sign and frequencies) in English

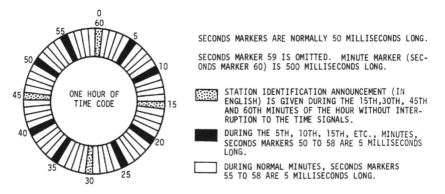

SECONDS MARKERS ARE NORMALLY 50 MILLISECONDS LONG.

SECONDS MARKER 59 IS OMITTED. MINUTE MARKER (SECONDS MARKER 60) IS 500 MILLISECONDS LONG.

STATION IDENTIFICATION ANNOUNCEMENT (IN ENGLISH) IS GIVEN DURING THE 15TH, 30TH, 45TH AND 60TH MINUTES OF THE HOUR WITHOUT INTERRUPTION TO THE TIME SIGNALS.

DURING THE 5TH, 10TH, 15TH, ETC., MINUTES, SECONDS MARKERS 50 TO 58 ARE 5 MILLISECONDS LONG.

DURING NORMAL MINUTES, SECONDS MARKERS 55 TO 58 ARE 5 MILLISECONDS LONG.

FIGURE 5.8. VNG TIME CODE.

and is approximately 30 seconds long, finishing approximately 10 seconds before each quarter hour. The speech is "notched" to allow the seconds markers to continue and has spectral components around 1000 Hz reduced to avoid erroneous operation of tuned-relay time signal receivers.

D. DUT1 Code

The difference between UTC and UT1 (DUT1) is given by the number of consecutive emphasized seconds markers, each one representing 0.1 second. DUT1 is positive when the first emphasized marker of a group is seconds marker 1 (i.e., the marker following the minute marker). DUT1 is negative when the first emphasized marker of a group is seconds marker 9. DUT1 is zero if no seconds markers are emphasized.

E. Accuracy

Carrier frequencies and 1000 Hz tone as emitted by VNG are maintained within 1 part in 10^{10} of Telecom Australia's standard of frequency (24 hour average value). As received, the accuracy may be degraded to the order of 1 part in 10^7 due to the ionosphere.

Time interval as emitted (i.e., elapsed time between any two seconds markers) has the same accuracy as the carrier frequencies except for intervals which include step adjustments.

Time of day as emitted is maintained within 100 microseconds of Telecom Australia's standard of Coordinated Universal Time,

UTC(ATC). As received, the signals may exhibit jitter of the order of 1 millisecond rms due to the ionosphere.

The Telecom Australia standard referred to above is based on cesium beam frequency and time standards at the Research Laboratories in Melbourne. This standard is maintained such that its value of frequency and time interval is within a few parts in 10^{12} of the international definition of time interval. Time-of-day signals generated by the Telecom Australia standard are maintained within about 100 microseconds of the international standard of UTC as determined by the BIH.

F. Frequency and Time Generating Equipment

The generating equipment is at Lyndhurst and includes precision quartz oscillators, frequency synthesizers, time code generators, announcing machines, and supervisory alarms.

A very low frequency signal, sent via landline from the Telecom Australia standard in Melbourne, controls the frequency of the operative quartz oscillator within 1 part in 10^{11}. Regular visits with a portable clock ensure that the VNG signals typically remain within 20 microseconds of UTC(ATC). General information is shown in Table 5.1.

5.1.9 ZUO

Radio station ZUO is operated by the Precise Physical Measurements Division of the National Physical Research Laboratory (NPRL) in Pretoria, South Africa. The signals are generated at NPRL and broadcast by the Post Office transmitting station at Olifantsfontein.

EXAMPLES

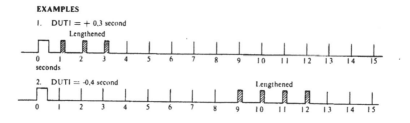

1. DUT1 = + 0.3 second

Lengthened

```
0    1    2    3    4    5    6    7    8    9   10   11   12   13   14   15
seconds
```

2. DUT1 = -0.4 second   Lengthened

```
0    1    2    3    4    5    6    7    8    9   10   11   12   13   14   15
```

FIGURE 5.9. DUT1 CODE ON ZUO.

A. Carrier Frequencies and Times of Transmission

ZUO transmits on 2.5 and 5 MHz. There is also a 100 MHz transmission which will not be discussed here. Carrier frequencies are normally derived from a cesium beam frequency standard at NPRL.

B. Standard Time Intervals and Time Signals

The transmitters are amplitude modulated by the time signals, the time difference (DUT1) code, and the Morse Code announcements. The signals consist of one pulse per second, each pulse consisting of 5 cycles of 1000 Hz. The first pulse in every minute is lengthened to 500 milliseconds.

Morse Code announcements are made during the minute preceding every fifth minute. They consist of the call sign ZUO (repeated three times) and the Universal Time at the next minute. The correct time is indicated by the beginning of a time pulse. Double-sideband modulation is employed. Modulation of the first half cycle of each pulse is positive.

C. Standard Audio Frequencies

No continuous frequencies are broadcast, but in many cases the interrupted 1000 Hz tone, which constitutes the time signal pulses, may be used as a standard frequency. The frequency of the Morse Code pulses is 600 Hz.

D. Accuracy

Time intervals and standard frequencies correspond to the adopted cesium beam frequency of 9 192 631 770 Hz. The frequency from which all other ZUO frequencies are derived can be relied on to be within 1 part in 10^{11} of the mean atomic frequency, as determined by the BIH. Standard time intervals have the same accuracy; i.e, 1×10^{-11}, with an additional uncertainty of ± 1 microsecond.

Time signals are Coordinated Universal Time (UTC) and are kept within 1 millisecond of the signals of other coordinated time stations. To keep UTC in approximate agreement with UT1, leap seconds are inserted when directed by the BIH. Inserted seconds are known as positive leap seconds, and omitted seconds as negative leap seconds.

E. DUT1 Code

The approximate value of the difference UT1 - UTC, to the nearest 10th of a second, is denoted DUT1; i.e., DUT1 = UT1 - UTC. DUT1 is indicated during the fifteen-second period following each minute mark. This is done by a group of lengthened second markers, the number in the group being the value of DUT1 in tenths of a second. If the group begins with the first second marker after the announced minute, DUT1 is positive. If the group begins with the ninth second marker, DUT1 is negative. If there are no lengthened second markers, DUT1 is zero. Examples are shown in figure 5.9.

76

TABLE 5.1.   STANDARD FREQUENCY AND TIME SIGNAL BROADCASTS IN THE HIGH FREQUENCY BAND

STATION	LOCATION LATITUDE LONGITUDE	POWER (kW)	ANTENNA	STANDARD FREQUENCIES USED CARRIER (MHz)	MODULATION (Hz)	TIMES OF UT TRANSMISSIONS	ACCURACY	DUT1 CODE	NOTES
ATA	NEW DELHI, INDIA 28° 33' N 77° 18' E	8	HORI-ZONTAL FOLDED DIPOLE	5.0; 10.0; 15.0	1; 1000	11 HRS/DAY MONDAY - SATURDAY 4 HRS/DAY ON 2ND SAT. OF MON. & SUN.	1×10^{-10}		SEE SECTION 5.1.3 FOR SPECIFIC DETAILS.
BPV	SHANGHAI, CHINA 39° 12' N 121° 26' E	5 10 15	OMNI-DIREC-TIONAL	5.0; 10.0; 15.0	1; 1000	19/30 (UTC) 5/30 (UT1)	1×10^{-10}	DIRECT EMMISSION OF UT1 SIGNAL	
BSF	TAIWAN, REP. OF CHINA	2		5.0		BETWEEN MIN 00-05, 10-15, 20-25, 30-35, 40-45, 50-55 FROM 0100-0900.		PULSE LENGTH-ENING TO 100 ms.	SECOND PULSES OF 5 ms DURATION. MIN MARKER IS PULSE OF 300 ms DURATION. DURING 29TH AND 59TH MIN, MORSE CODE AND CHINESE VOICE ANNOUNCEMENT OF TIME.
CHU	OTTAWA, CANADA 45° 18' N 75° 45' W	3 10 3	OMNI-DIRECT-IONAL	3.330; 7.335; 14.670		CONTINUOUS	5×10^{-12}	CCIR CODE: SPLIT PULSE	SEE SECTION 5.1.2 FOR SPECIFIC DETAILS
DAM	ELMSHORN, GERMANY, F.R. 53° 46' N 9° 40' E	10 15 5 10 5 15		8.6385 16.9804} 4.2650 8.6385} 6.4755 12.7635}		1155-1206 2355-2406 FROM 21 SEPT. TO 20 MAR. 2355-2406 FROM 21 MAR. TO 20 SEPT.		DOUBLE PULSE	NEW S1, THEN SECOND PULSES FROM MINS 0.5 TO 6.0. MIN PULSE PROLONGED. *DUT1 TRANSMITTED AFTER MIN PULSES 1 TO 5.
FFH	PARIS, FRANCE 48° 33' N 2° 34' E	5	VERTICAL DIPOLE	2.5		CONTINUOUS	2×10^{-11}	CCIR CODE: PULSE LENGTH-ENING TO 100 ms.	SEC PULSES OF 5 CYCLES OF 1 kHz MODULATION. MINUTE PULSES PROLONGED TO 500 ms.

STATION	LOCATION LATITUDE LONGITUDE	POWER (kW)	ANTENNA	CARRIER (MHz)	MODULATION (Hz)	TIMES OF UT TRANSMISSIONS	ACCURACY	DUT1 CODE	NOTES
IAM	ROME, ITALY 41° 47' N 12° 27' E	1	VERTICAL DIPOLE λ/4	5.0	1; 1000	CONTINUOUS	5×10^{-11}	CCIR CODE: DOUBLE PULSE	SEE SECTION 5.1.4 FOR SPECIFIC DETAILS.
IBF	TORINO, ITALY 45° 2' N 7° 46' E	5	VERTICAL DIPOLE λ/4	5.0	1	CONTINUOUS	1×10^{-11}	CCIR CODE: DOUBLE PULSE	SEE SECTION 5.1.5 FOR SPECIFIC DETAILS.
JJY	SANWA, SASHIMA, IBARAKI, JAPAN 36° 11' N 139° 51' E	2	VERTICAL λ/4 (2.5 MHz) HORIZ. λ/2 DIPOLE (5 MHz) VERTICAL λ/2 DIPOLE (10 & 15 MHz)	2.5; 5.0; 10.0; 15.0	1; 1000	CONTINUOUS	1×10^{-11}	CCIR CODE: PULSE LENGTHENING	SEE SECTION 5.1.6 FOR SPECIFIC DETAILS.
LOL	BUENOS AIRES, ARGENTINA 34° 37' S 58° 21' W	2	HORIZONTAL 3-WIRE FOLDED DIPOLE	5.0; 10.0;	1; 440; 1000	CONTINUOUS	2×10^{-11}	CCIR CODE: PULSE LENGTHENING	SEC PULSES OF 5 CYCLES OF 1000 Hz MODULATION. SEC 59 IS OMITTED. ANNOUNCEMENT OF HRS & MINS EVERY 5 MINS, FOLLOWED BY 3 MINS OF 1000 Hz & 440 Hz MODULATION.
MSF	RUGBY, UNITED KINGDOM 52° 22' N 1° 11' W	5	HORIZONTAL QUADRANT DIPOLES; (VERTICAL MONOPOLE, 2.5 MHz)	2.5; 5.0; 10.0	1	BETWEEN MIN 0-5, 10-15, 20-25, 30-35, 40-45, 50-55	2×10^{-12}	CCIR CODE: DOUBLE PULSE	SEC PULSES OF 5 CYCLES OF 1 kHz MODULATION. MIN PULSES ARE PROLONGED. CALL SIGN IN VOICE & MORSE CODE.
OMA	LIBLICE, CZECHOSLOVAKIA 50° 04' N 14° 53' E	1	T	2.5	1; 1000	BETWEEN MIN. 5-15, 25-30, 35-40, 50-60 OF EVERY HR. EXCEPT 0500-1100 ON 1ST WED. OF EACH MONTH.	5×10^{-11}	NO TRANSMISSION OF DUT1.	SEE SECTION 5.1.7 FOR SPECIFIC DETAILS.

		Power	Antenna	Frequency (MHz)		DURING 5 MINS PRECEDING			MIN PULSES ARE PROLONGED.
PPR	RIO DE JANEIRO, BRAZIL 22° 59' S 43° 11' W			4.35 8.634 13.105 17.1944		0130, 1430, 2130.			
RCH	TASHKENT, USSR 41° 19' N 69° 15' E	1	HORIZONTAL DIPOLE	2.5	1; 10	BETWEEN MIN. 5-10, 15-20, 35-40, 45-50, FROM 0000-0350, 0535-0930, 1015-1330, 1415-2400.	2×10^{-10}	*MORSE CODE	MINUTE PULSES PROLONGED TO 500 ms. *DUT1 + dUT1 EACH HR BETWEEN MIN. 51 & 52.
RID	IRKUTSK, USSR 52° 46' N 103° 39' E	1	HORIZONTAL DIPOLE	5.004 10.004 15.004	1; 10	BETWEEN MIN 5-10, 15-20, 25-30, 51-60, FROM 0000-0110, 1351-2400. BETWEEN MIN 5-10, 15-20, 25-30, 51-60, FROM 0151-1310.	5×10^{-11}	*MORSE CODE	MIN PULSES PROLONGED. *DUT1 + dUT1 EACH HR BETWEEN MIN 31 & 31.
RIM	TASHKENT, USSR 41° 19' N 69° 15' E	1	HORIZONTAL DIPOLE	5.0; 10.0	1; 10	BETWEEN MIN 15-20, 25-30, 35-40, 45-50, FROM 0000-0130, 0215-0350, 1815-2400. BETWEEN MIN 15-20, 25-30, 35-40, 45-50, FROM 0535-0930, 1015-1330, 1415-1730		*MORSE CODE	MIN PULSES PROLONGED TO 50 ms. *DUT1 + dUT1 EACH HR BETWEEN MIN 51 & 52.
RTA	NOVOSIBIRSK, USSR 55° 46' N 82° 58' E	5	HORIZONTAL DIPOLE	10.0; 15.0	1; 10	BETWEEN MIN. 5-10, 15-20, 25-29, 35-39, FROM 0130-0300, 1750-2400. BETWEEN MIN. 30-35, 41-45, 50-60, FROM 0350-1700	5×10^{-11}	*MORSE CODE	MIN. PULSES PROLONGED *DUT1 + dUT1 EACH HR BETWEEN MIN 11 & 12.
RMM	MOSCOW, USSR 55° 19' N 38° 41' E	5 5 8	HORIZONTAL DIPOLE	4.996 9.996 14.996	1; 10	39 IN EACH 60	5×10^{-11}	*MORSE CODE	MIN. PULSES PROLONGED TO 500 ms. *DUT1 + dUT1 EACH HR BETWEEN MIN 11 & 12.
VNG	LYNDHURST, VICTORIA, AUSTRALIA 38° 03' S 145° 16' E	10	OMNI-DIRECTIONAL	4.5 7.5 12.5	1; 1000*	0945-2130 (4.5 MHz) 2245-2230 (7.5 MHz) 2145-0930 (12.5 MHz)	1×10^{-10}	*CCIR CODE	*SEE SECTION 5.1.8 FOR SPECIFIC DETAILS.

STATION	LOCATION LATITUDE LONGITUDE	POWER (kW)	ANTENNA	STANDARD FREQUENCIES USED		TIMES OF UT TRANSMISSIONS	ACCURACY	DUT1 CODE	NOTES
				CARRIER (MHz)	MODULATION (Hz)				
YVTO	CARACAS, VENEZUELA 10° 30' N 66° 56' W			6.1		CONTINUOUS			SEC PULSES OF 1 kHz MODULATION WITH 100 ms DURATION. MIN IDENTIFIED BY 800 Hz TONE OF 500 ms DURATION. SEC 30 IS OMITTED. VOICE ID OF HR, MIN, SEC, EACH MIN BETWEEN SECS 52 & 57.
WWV	FT. COLLINS, COLORADO 40° 41' N 105° 02' W	2.5 10.0 10.0 2.5	VERTICAL λ/2 DIPOLE ARRAYS	2.5; 5.0; 10.0; 15.0; 20.0	1; 440; 500; 600	CONTINUOUS	1 x 10⁻¹¹	CCIR CODE: DOUBLE PULSE	SEE SECTION 5.1.1 FOR SPECIFIC DETAILS.
WWVH	KEKAHA, KAUAI, HAWAII 21° 59' N 159° 46' W	5 10 10	PHASED VERTI-CAL HALF-WAVE DIPOLE ARRAYS	2.5 5.0 10.0 15.0	1; 440; 500; 600	CONTINUOUS	1 x 10⁻¹¹	CCIR CODE: DOUBLE PULSE	SEE SECTION 5.1.1 FOR SPECIFIC DETAILS.
ZUO	OLIFANTSFONTEIN, REPUBLIC OF SOUTH AFRICA 25° 58' S 28° 14' E	4	VERTICAL MONOPOLE	2.5; 5.0	1	1800-0400 (2.5 MHz) CONTINUOUS (5 MHz)	1 x 10⁻¹¹	PULSE LENGTH-ENING	SEE SECTION 5.1.9 FOR SPECIFIC DETAILS.

5.2 RECEIVER SELECTION

This and the following sections discuss the use of HF broadcasts for frequency and time calibrations. For convenience, we will refer to mainly to WWV; however, the discussions also apply to other HF broadcasts.

The reception and use of shortwaves can be quite tricky. However, for many locations, simple receivers and antennas work well most of the time. Also, there are many separate signals available from WWV and WWVH. By choosing from among the many combinations of frequencies and distances involved, a solution usually can be found for any problem.

For those users who must have the very best signal most of the time, this section discusses several methods available to solve reception problems. The user is also encouraged to use other references on HF propagation, antennas, and receiver selection and operation.

Almost any shortwave receiver can be used to receive HF signals for time and frequency calibrations. However, for accurate time and frequency calibrations, especially in noisy locations, it is important to have a good antenna-ground system and a receiver with optimum sensitivity, selectivity, image rejection, and frequency or phase stability. A block diagram of a typical high-performance receiver is shown in figure 5.10.

The first requirement of a good receiver is sensitivity. Therefore, a tuned RF amplifier is desirable because it increases sensitivity. The next requirement is selectivity. This is the ability of a receiver to reject nearby signals that interfere with the desired signal. Selectivity is achieved by heterodyning (translating) the incoming signal to a much lower fixed frequency (usually 455 kHz) called the intermediate frequency (IF). The IF amplifiers are fixed-tuned for maximum gain with a relatively narrow frequency pass band. Often, either crystal or mechanical filters are used in the IF stages to provide an even narrower pass band of frequencies to reject interference.

The third feature of a good receiver is the ability to reject interference from an undesired signal at what is called the image frequency. Since the IF signal is the difference between the local oscillator frequency and the incoming signal, there are always two different incoming signals that can produce the same IF signal. One is above and the other below the oscillator frequency.

For example, a receiver with an intermediate frequency of 455 kHz tuned to WWV at 10.0 MHz would have an oscillator frequency of 10.455 MHz (10.455 - 10.0 = 455). However, if another signal with a frequency of approximately 10.910 MHz is present, another "difference" signal of 455 kHz is produced (10.910 - 10.455 = 455). Thus, if two signals are separated by twice the intermediate frequency and the oscillator frequency falls between the two signals, an unwanted image signal results. Table 5.2 lists various dial settings and corresponding oscillator frequencies that can receive WWV signals as image frequencies if the receiver IF is 455 kHz.

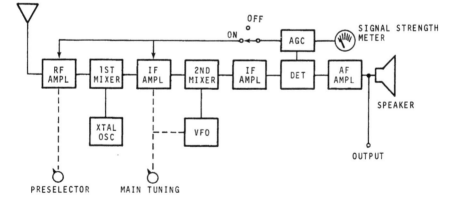

FIGURE 5.10. BLOCK DIAGRAM OF A TYPICAL HIGH-PERFORMANCE HF RECEIVER.

TABLE 5.2 IMAGE FREQUENCIES FOR WWV/WWVH

OSCILLATOR FREQUENCY (MHz)	IMAGE FREQUENCY (MHz)	DIAL FREQUENCY (MHz)	OSCILLATOR FREQUENCY (MHz)	IMAGE FREQUENCY (MHz)	DIAL FREQUENCY (MHz)
2.045	1.590	2.500	14.545	14.090	15.000
2.955	3.410	2.500	15.455	15.910	15.000
4.545	4.090	5.000	19.545	19.090	20.000
5.455	5.910	5.000	20.455	20.910	20.000
9.545	9.090	10.000			
10.455	10.910	10.000			

Images occur because the IF signal is very low in frequency (i.e., 455 kHz) relative to the high frequency of the incoming signal (i.e., 10 MHz). Thus, the RF amplifier may have insufficient selectivity to reject interference which is relatively close to the desired signal. In the above example, the unwanted signal is only 0.910 MHz away from 10 MHz and, if strong, might easily pass through the RF stage into the mixer and produce an image IF signal.

To solve this problem, a high quality receiver usually employs double conversion or two mixer stages. It heterodynes in two steps to make image rejection easier. The first mixer stage converts the incoming signal to a high intermediate frequency (usually greater than 1 MHz). Any images produced will be separated by more than 2 MHz from the desired signal. For example, if a receiver has a first mixer IF signal of 1.5 MHz, the image signal will be twice the IF signal or 3.0 MHz above the desired signal. Since 3.0 MHz is now relatively large compared to the incoming signal, the RF stage pass band is usually selective enough to reject the undesired images. For the required additional selectivity and gain, a second mixer stage is added to produce a lower second IF frequency, thus solving the image problem.

A local crystal oscillator can be used in these conversion processes to provide tuning stability. Other information regarding well-designed receivers can be found in such reference material as the Radio Amateur's Handbook.

If all the above seems too confusing, don't worry. Many small, inexpensive imported and domestic receivers work very well. In fact, several receivers have special buttons that you can push and the receiver automatically tunes itself to WWV. If timing is a special problem in your home or business, of course check out the receiver very carefully before you purchase it. Image and interference

signals come and go so they may not persist long enough to be a serious problem. The single best option to look for in a receiver is the provision for an external antenna. This may be just a short wire on the windowsill or an elaborate rooftop installation. Fancy knobs and adjustments won't help much if you can't get a good signal.

5.3 CHOICE OF ANTENNAS AND SIGNAL CHARACTERISTICS

A good antenna is important if you want strong, noise-free signals. However, follow the basic rule of "using what you have available." Many receivers with built-in antennas work quite satisfactorily. Often, adding a short piece of wire to the existing receiver antenna will bring in a stronger signal that can be heard well enough to set your watch.

But for the serious listener or the metrology lab that depends on WWV or other HF signals for calibrations, something more is needed. In the material that follows, the rules for antenna selection and installation are discussed with respect to choices available. Companies that sell WWV receivers often offer antennas. These have been proven by use and the instructions included with them will help a lot in getting a good usable signal for frequency and time calibrations.

The signal power as received at great distances from HF stations may be relatively weak. As the distance is increased, the signal decreases in strength so an antenna with maximum efficiency produces the best results. Radio waves arrive at different vertical angles, called the wave or radiation angle. The wave angle of an arriving signal depends on the distance between the transmitting and receiving stations and also on the height of the ionosphere. The closer the receiving station and the higher the ionosphere, the larger the wave angle. Therefore, an antenna should be selected and oriented that favors the wave angle (elevation) as well as the direction (azimuth). Determining wave angles is discussed more fully in Section 5.6.

Many HF stations transmit on a number of frequencies. Because of changes in ionospheric conditions, which sometimes adversely affect the signal transmissions, most receivers are not able to pick up the signals on all frequencies at all times in all locations. Except during times of severe magnetic disturbances, however--which make radio transmissions almost impossible--users should be able to receive the signal on at least one of the broadcast frequencies. As a general rule, frequencies above 10 MHz provide the best daytime reception while the lower frequencies are best for nighttime reception.

An all-band antenna capable of covering the entire HF band is commercially available. These are used for around-the-clock communication purposes where the entire HF band must be used for maximum reliability. Such an antenna is very large and cumbersome as well as expensive. Depending on location, an antenna and a receiver capable of receiving two or more frequencies may be required. Any of the different WWV frequencies transmitted may be received depending upon distance, time of day, month, season, and sunspot cycle. Each frequency will be discussed with the type of antenna most suitable for its reception. Although WWV and WWVH are not currently broadcasting on 25 MHz, this frequency is also discussed.

In the following discussion, mention is made of sunspot cycles. These cycles occur with a periodicity of about every eleven years. Propagation characteristics at high frequencies are greatly affected by the sunspot cycles. One solution to the problem is to select another frequency that will work and to ignore the ones that won't. For those users who want to understand the propagation effects and to counteract the coming sunspot cycles by antenna selection and placement, we recommend further reading on this subject. Information on sunspots is available from the Space Environment Lab, NOAA, U. S. Department of Commerce. Minimums in the sunspot cycle occurred in 1954 and 1965. The corresponding maximums were in 1958 and 1969, and another will probably occur in 1980.

ANTENNAS FOR THE 2.5 TO 4 MHz RANGE.

Signals at these frequencies have a very short range during the day because of groundwave propagation limitations. Daytime use is limited to locations within one to two hundred miles from the transmitter. They become useful at night, however, especially during the winter season in the higher latitudes where longer nights prevail. Reception is then possible over distances of several thousand miles.

The vertical quarter-wavelength monopole antenna has a radiation pattern that favors reception at low wave angles and is very effective in receiving long distance skywave signals normally arriving at angles of 20 degrees or less. This antenna is not effective for reception of short-range skywave signals with large wave angles, but it is useful for receiving weak groundwave signals.

For nighttime reception on paths up to several thousand miles, the skywave is predominant and a horizontal half-wavelength antenna is recommended. The antenna should be located a quarter-wavelength or higher above bround and separated from possible interfering reflecting obstacles. The quarter-wavelength vertical antenna and the half-wavelength horizontal dipole antenna are shown in figure 5.11. To get antenna dimensions for CHU or other frequencies, refer to antenna manuals such as those published by the American Radio Relay League.

ANTENNAS FOR THE 4 TO 5.0 MHz RANGE.

These frequencies can be received at greater distances than the 2.5 MHz frequency throughout the day or night, especially during minimum sunspot cycle. Reception is possible up to 1000 miles under ideal conditions, but under normal conditions, daytime propagation conditions limit its useful range. Therefore, reception is usually limited to less than 1000 miles during the day. During the night, 5 MHz is a very useful frequency for long-range reception except during maximum sunspot cycle. It is excellent during early dawn and early evening in the winter months when the signal path is in darkness. Horizontal dipole antennas are a good choice for this band. Users should also consider the possibility of a multiband antenna.

ANTENNAS FOR THE 7 TO 10.0 MHz RANGE.

At these frequencies, reception over great distances is possible during the day or night. They can be classified as intermediate frequencies but again, they are dependent upon the sunspot cycle. During minimum sunspot cycle, great distances can be covered during the day when higher frequencies cannot be received. During maximum sunspot cycle, these are probably the best frequencies to use during the night when the lower frequencies cannot be heard. During maximum sunspot cycle, reception is limited to short distances

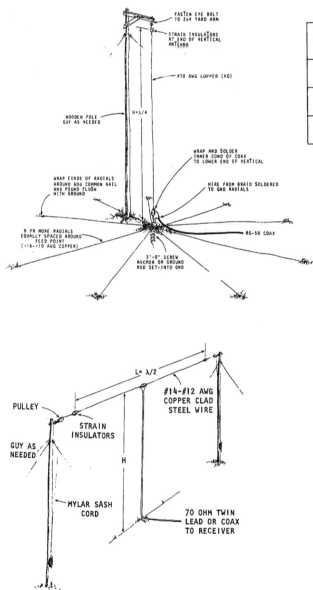

FREQUENCY MHz	1/4 λ VERTICAL ANTENNA HEIGHT
2.5	28.37 m 94'- 7"
5.0	14.05 m 46'-10"
10.0	7.05 m 23'- 6"

FIGURE 5.11. QUARTER-WAVELENGTH ANTENNA AND HORIZONTAL HALF-WAVELENGTH ANTENNA.

uring the day with limitations comparable to
hose noted for 5.0 MHz. These frequencies
lso provide daytime reception at fairly close
ange, 200 miles or so, and can be used when 5
Hz reception is poor. The half-wavelength
orizontal antenna should be selected for
hort distances. A quarter-wavelength verti-
al antenna is suitable at greater distances
see figure 5.11).

ANTENNAS FOR THE 14 TO 15 MHz RANGE.

These frequencies are most favorable for
ong range, daytime reception. They are not
sable for short-range reception except during
eriods of sunspot maximum. However, for long
ange reception, they are the most favored
requencies during both sunspot cycle condi-
ions. Under average conditions, the maximum
ave angle is limited to 30 degrees or less
epending on the density of the ionized layer.

During maximum sunspot cycle, reception
ay even be possible during the night in some
ocations. During minimum sunspot cycle, they
re useful only during the daylight hours and
awn and dusk periods. Both horizontal and
ertical antennas are suitable.

ANTENNAS FOR 20 MHz.

The 20 MHz frequency is normally the best
to use for daytime reception and will be optimum at either noon or a few hours past noon.
Signals at this frequency arrive at very low
wave angles and are useful only for long distance reception. During minimum sunspot
cycles, reception is poor but improves during
the winter. During maximum sunspot cycles,
the reception is excellent at night and during
the day. The vertical dipole that favors low
wave angle radiation has been used at this
frequency with favorable results. Construction details for a 20 MHz antenna are shown in
figure 5.12.

ANTENNAS FOR 25 MHz.

The 25 MHz frequency is best during daylight hours except during the summer. Reception is especially good during maximum sunspot
cycle and very poor during minimum sunspot
cycle. It is used only for long distance
reception because a low radiation angle is
required for the signal to be returned back to
the earth from the ionosphere. Design details
of a different vertical antenna for use at 25
MHz are also shown in figure 5.12.

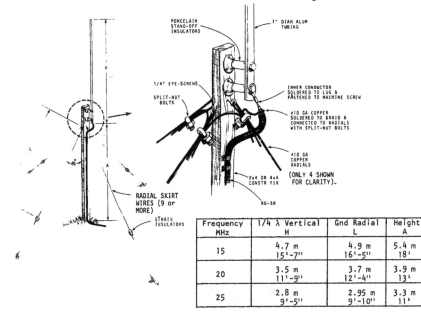

Frequency MHz	1/4 λ Vertical H	Gnd Radial L	Height A
15	4.7 m 15'-7"	4.9 m 16'-5"	5.4 m 18'
20	3.5 m 11'-9"	3.7 m 12'-4"	3.9 m 13'
25	2.8 m 9'-5"	2.95 m 9'-10"	3.3 m 11'

FIGURE 5.12. MODIFIED HALF-WAVELENGTH VERTICAL ANTENNA FOR USE AT 15, 20, AND 25 MHz.

VERTICAL ANTENNA CONSTRUCTION.

The vertical antenna which is favorable to low wave-angle reception is preferable to the horizontal half-wavelength antenna and can readily be constructed as shown in figure 5.12. The dimension of the ground radials and orientation as shown should be used to yield approximately a 50 ohm antenna impedance. For a 70 - 90 ohm antenna, the half-wavelength dipole mounted vertically will yield approximately the correct impedance. In order to prevent interaction between the feed line and the lower half of the dipole, which disturbs the radiation pattern, extend the feed line horizontally outward several feet from the antenna before dropping it vertically to the ground.

5.4 USE OF HF BROADCASTS FOR TIME CALIBRATIONS

Now that we have a receiver and an antenna, we can proceed to make time calibrations. If your accuracy requirements are low, you can use the voice time-of-day announcements. However, if you want higher accuracy, you will have to measure the seconds ticks or decode the WWV/WWVH time code.

5.4.1 TIME-OF-DAY ANNOUNCEMENTS

Time of day is available from many sources in the United States and Canada. Radio and television stations mention the time frequently and, in fact, use the time of day to schedule their own operations. The telephone companies offer time-of-day services in most locations. But where does the time come from?

Most time-of-day services in the U. S. start at the National Bureau of Standards. An NBS telephone service is available by calling (303) 499-7111. In addition, WWV and WWVH broadcast voice time-of-day announcements once each minute. This is also the case for CHU where the time announcements are given alternately in French and English. Using the WWV voice announcement and the tone following the words, "At the tone, XXXX hours XXXX minutes Coordinated Universal Time," a person can check a wall clock or wristwatch to within a fraction of a second. The UTC time that ·is announced can be converted to local time by using the map in Chapter 2.

5.4.2 USING THE SECONDS TICKS

For accurate time recovery, we cannot just listen. We must rely on electronic equipment. By using an electronic oscilloscope to see the ticks, the user can set clocks with much greater accuracy than is possible by only listening to the ticks. An oscilloscope is a necessity when dealing with noisy signals that fade.

The following describes ways to get time of day with resolutions better than 1 millisecond (1000 microseconds) by using an oscilloscope. Under very favorable conditions, it is possible to establish time synchronization to about 100 microseconds. In each case, path and equipment time delay corrections are necessary for accurate results.

In order to achieve the best results in any of these measurement methods, the following guidelines are recommended:

1. Make your measurements at the same time every day.

2. Avoid twilight hours when the ionosphere is the least stable.

3. Use the highest carrier frequency which provides consistently good reception.

4. Look at the received signals on the oscilloscope for a few minutes to judge the stability of propagation conditions and select that part of the timing waveform that is most consistent.

Since receiver and propagation time delays are so important, they will be discussed before we go into the details of performing timing calibrations.

A. Receiver Time Delay Measurements

For accuracy in time-setting, the receiver delay must be known. Typical delays in receiver circuits are usually less than a millisecond. However, they should be known to the user who wants the very best timing accuracy. The actual delay in the receiver will vary with tuning and knob settings.

To measure the receiver time delay, the following equipment is required: An oscilloscope with accurately calibrated, externally-triggered time base; an HF signal generator; and an audio signal generator. The equipment is connected as shown in figure 5.13.

The main tuning dial of the receiver should be set at the exact position for receiving WWV/WWVH signals. This is because the receiver delay time varies with slightly different receiver dial positions. Therefore, the receiver tuning should be set and marked

86

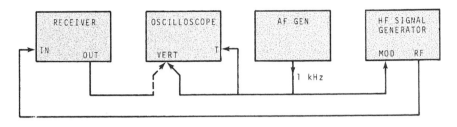

FIGURE 5.13. EQUIPMENT SETUP FOR RECEIVER TIME DELAY MEASUREMENTS.

where the maximum signal is received. The frequency of the HF signal generator is then adjusted for peak receiver output.

The audio signal generator is set to a 1 kHz output frequency. A high accuracy 1 kHz signal is not required. The HF generator is externally modulated by the 1 kHz signal. The oscilloscope sweep rate is set to 100 microseconds/division with positive external triggering from the 1 kHz signal. The vertical amplifier gain is set high for a large vertical deflection. The vertical position control is adjusted for zero baseline with no input signal.

Initially, the 1 kHz signal generator is connected to the vertical input of the oscilloscope. The trigger level control is adjusted so that the trace crosses over at the horizontal center line at the left. The horizontal position control can be adjusted so that the signal crosses over the first division on the left as shown in figure 5.14. The crossover point of the undelayed signal will serve as the zero delay reference point.

Without touching any of the oscilloscope controls, the 1 kHz signal is disconnected from the vertical input and replaced with the delayed 1 kHz signal from the receiver output. Since a receiver delay is almost always less than 1 millisecond (1 cycle of a 1 kHz signal), there is little chance of ambiguity on which cycle to measure. The delay is equal to the sweep time from the reference undelayed crossover to the first delayed crossover point.

If the delayed signal is found to be of opposite phase to the reference 1 kHz signal, the receiver has an inverted output signal. However, the receiver delay time remains unchanged and the only difference will be that the output seconds pulse will be inverted with a negative leading edge. For receiver delays of less than 500 microseconds, the sweep rate can be set at 50 microseconds/division.

This technique is a way of producing a local signal that approximates the timing signal. It uses the same frequency on the dial and uses the same tone frequency. A two-channel scope is very useful since you can then display the tone before it enters the RF signal generator and after it comes out of the receiver. A study of the block diagram (fig. 5.13) will no doubt suggest other variations on this technique to the user.

B. Time Delay Over the Radio Path

In order to make a calibration, a user must know the path delay from the transmitter to his receiver antenna. This is often difficult due to the fact that several "modes" of propagation are possible at high frequencies. A discussion of propagation delay calculations can be found in section 5.6.

For time setting of highest accuracy, a user would have to actually measure the path delay using a portable clock. This is an involved and costly operation that few users would deem justified for their needs. However, once the path delay is determined, it

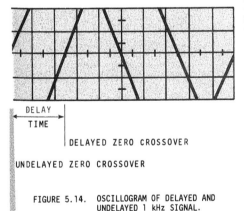

DELAY
TIME

DELAYED ZERO CROSSOVER

UNDELAYED ZERO CROSSOVER

FIGURE 5.14. OSCILLOGRAM OF DELAYED AND UNDELAYED 1 kHz SIGNAL.

87

need not change appreciably if certain precautions are observed. For timing accuracies that approach 100 microseconds, the location and routing of antennas and lead-ins, are important as are receiver settings. The general rule is to "not change anything". If changes must be made, careful records of readings before and after will help in re-establishing the system with a correct delay value.

In fact, any user of time signals at high levels of accuracy must take extreme care in how the equipment is adjusted, maintained, and operated. Typical users estimate their path delays and obtain results with accuracies approaching one millisecond. The above paragraphs are intended for those users who want to obtain the very best accuracy. They are not typical for the large percentage of users. It also goes without saying that a wealth of information can be obtained by reading the literature supplied with the timing equipment.

C. Using an Adjustable Clock to Trigger the Oscilloscope

This method, sometimes called the Direct Trigger Method, is the simplest and requires an oscilloscope with an external sweep trigger and accurately calibrated time base, and a receiver with electronic audio output. The clock must be adjustable so its 1 pps output can be advanced or retarded. Equipment connection is shown in figure 5.15.

A local clock pulse at 1 pps second rate is used to trigger the oscilloscope sweep. At some time interval later during the sweep, the seconds pulse appears on the display as shown. The time interval from the start of the sweep to the point where the tick appears is the total time difference between the local clock and the transmitting station. By subtracting the propagation and receiver time delays from the measured value, the local clock time error can be determined. The equation to determine time error at a receiving location is:

$$\text{Time Error} = t_r - t_t = TD - (TD_p + TD_r),$$

where: t_r = time at receiving station

t_t = time at transmitting station (WWV, WWVH)

TD = total time difference (measured by oscilloscope display)

TD_p = propagation path time delay (computed)

TD_r = receiver time delay (measured)

NOTE: The units (usually milliseconds) should be the same.

The receiver is tuned to the station and the oscilloscope sweep rate set at 0.1 s/division. Listening to the broadcast will help judge the quality of reception and fading. The tick will typically appear as shown in figure 5.16. CHU signals can be used as well (see fig. 5.17). If the tick is one division or more from the left side of the scope display, the time of the local clock is corrected until the tick falls within the first division from the left side. If the local time tick is late, the received tick will be heard before the sweep starts. If this is the case, the local clock should be adjusted until the tick appears. After the local seconds pulse has been properly adjusted and appears within the first division (0.1 second in time), the sweep rate is increased to say, 5 ms/division. Using this greater resolution, the local clock is adjusted until the leading edge of the received pulse starts at a time equal to the propagation delay time plus the receiver delay time after the trigger as shown in figure 5.18.

The sweep rate should be increased to the highest rate possible without allowing the total sweep time to become faster than the combined propagation and receiver delay time

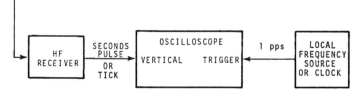

FIGURE 5.15. BLOCK DIAGRAM OF EQUIPMENT CONNECTION FOR DIRECT TRIGGER METHOD OF TIME SYNCHRONIZATION.

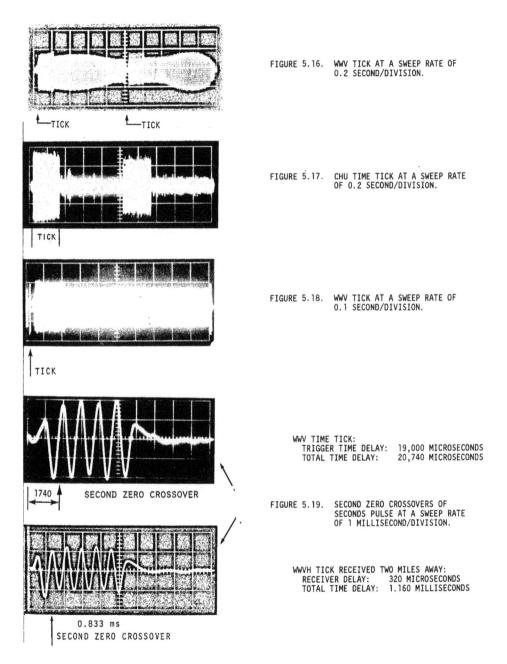

—TICK   —TICK

FIGURE 5.16.   WWV TICK AT A SWEEP RATE OF
               0.2 SECOND/DIVISION.

TICK

FIGURE 5.17.   CHU TIME TICK AT A SWEEP RATE
               OF 0.2 SECOND/DIVISION.

TICK

FIGURE 5.18.   WWV TICK AT A SWEEP RATE OF
               0.1 SECOND/DIVISION.

WWV TIME TICK:
    TRIGGER TIME DELAY: 19,000 MICROSECONDS
    TOTAL TIME DELAY: 20,740 MICROSECONDS

1740   SECOND ZERO CROSSOVER

FIGURE 5.19.   SECOND ZERO CROSSOVERS OF
               SECONDS PULSE AT A SWEEP RATE
               OF 1 MILLISECOND/DIVISION.

WWVH TICK RECEIVED TWO MILES AWAY:
    RECEIVER DELAY: 320 MICROSECONDS
    TOTAL TIME DELAY: 1.160 MILLISECONDS

0.833 ms
SECOND ZERO CROSSOVER

89

less the 5 milliseconds to allow for the
length of the received seconds pulse.

With a sweep rate of 1 ms/division, for
example, greater resolution can be realized by
measuring the second zero crossover point of
the 5 ms received tick. Although the leading
edge of the seconds pulse as broadcast from
these stations is "on time", coincident with
the NBS or CHU standard, it is difficult to
measure because of the slow rise time at the
beginning of the burst and distortion due to
propagation. For this reason, the second zero
crossover should be used. The second zero
crossover of the WWV or CHU pulse is delayed
exactly 1000 microseconds and the WWVH crossover is delayed 833 microseconds as shown in
figure 5.19. This is called the cycle
correction.

At a sweep rate of 1 ms/division, any
changes in arrival time (jitter) are readily
apparent. After watching the pulses for a
period of a minute or two, select a cycle that
is undistorted and relatively larger in amplitude. In determining the time at a receiving
location, include the delay of the chosen zero
crossover point, then add the cycle correction
to the propagation and equipment delay using
the following relationship:

Time Error = $t_r - t_t$

$\quad$ = TD $-(TD_p + TD_r +$ cycle correction$)$

Where cycle correction time

$\quad$ = 1000 microseconds per cycle (WWV or CHU)

$\quad$ = 833 microseconds per cycle (WWVH)

As an example, assume an operator at a
distant receiver location is interested in
comparing his time to that of WWVH. The propagation and receiver time delays were measured as 11.7 milliseconds and 300 microseconds respectively. Since the total delay is
12.0 milliseconds (11.7 ms + 0.3 ms), the
oscilloscope sweep rate was set at 2 ms/
division for a total sweep time of 20 ms--
slightly greater than the propagation delay +
receiver delay + 5 ms total. The second zero
crossover of the tick was observed and measured 12.5 ms after the sweep was triggered by
the WWVH clock.

From these data, the time at the receiver
site was calculated as -0.333 ms with respect
to the WWVH time as broadcast, or:

Time Error = $t_r - t_t$

$\quad$ = 12.5 - (11.7 + 0.3 + 0.833)

$\quad$ = -0.333 ms, or -333 microseconds,

Where:$\quad$ TD $= 12.5$ ms

$\quad\quad\quad$ $TD_p \equiv 11.7$ ms

$\quad\quad\quad$ $TD_r = 0.3$ ms

Note that if a receiving station is
located at a distance greater than 3,000 km
(1863 miles) from the transmitter, the propagation time delay will exceed 10 ms. (The
radio path delay works out to be about 5
microseconds per mile. At 1863 miles, the
delay would be at least 9.315 ms. It is
usually greater than this due to the fact the
HF radio signals bounce off the ionosphere.)
This forces use of a scope sweep time of 2 ms/
division and lowers the measurement resolution. The next section describes a method of
measurement to overcome this difficulty.

D. Delayed Triggering: An Alternate Method
   that Doesn't Change the Clock Output

To improve the resolution of measurement,
the oscilloscope sweep must be operated as
fast as possible. The user does have an
option: He can generate a trigger pulse independent of his clock. He then positions the
pulse for maximum sweep speed and makes his
measurement. But then he must measure the
difference between his clock and the trigger
pulse. Note: This can be accomplished by
using an oscilloscope with a delayed sweep
circuit built in or with an outboard trigger
generator. The latter method is discussed
here, but the delayed sweep scope could be
used as discussed in Chapter 4. Reference to
the instrument manual will aid in using that
technique.

On a typical digital delay generator, a
delay dial indicates the delay between the
input local clock tick and the output trigger
pulse. If the user already has a variable
rate divider to produce delayed pulses, a time
interval counter can be used instead of the
delay generator. In either case, the amount
of trigger delay must be accounted for in
measuring the total time delay (TD) of the
received tick with respect to the local master
clock.

Measurements should be made at the same
time every day (within ten minutes) for consistent results. A time of day should be
selected when it is approximately noon midway
between the transmitting station and the receiver's location. For night measurements, a
time should be chosen when the midpoint is
near midnight. Measurements should not be
made near twilight.

The equipment can be connected as shown
in figure 5.20. A commercially available

90

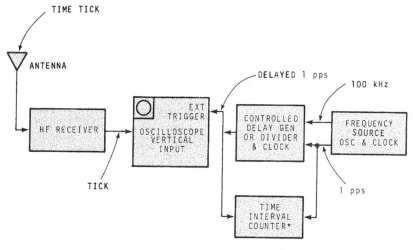

TIME TICK

ANTENNA

DELAYED 1 pps

100 kHz

*SEE TEXT

FIGURE 5.20. EQUIPMENT SETUP FOR DELAYED TRIGGER METHOD OF TIME SYNCHRONIZATION.

frequency divider and clock can be used in place of the controlled delay generator. A time interval counter is then used to measure the output of the delayed clock to the master clock. The output of the delayed clock is used to trigger the oscilloscope. The initial procedures described in the previous method also apply to this method and, therefore, should be referred to in setting time with WWV/WWVH.

With the oscilloscope sweep adjusted to 1 ms/division, the trigger pulse should be

delayed by an amount equal to the propagation delay in milliseconds. Initially, any fractional milliseconds in the delay can be neglected. The sweep should be adjusted so that it begins exactly at the left end of the horizontal graticule and is vertically centered.

The second zero crossover point of the tick (figures 5.21, 5.22) should be observed and carefully measured. With the sweep at 1 ms/division, the delay of the second zero crossover on the oscilloscope is measured to

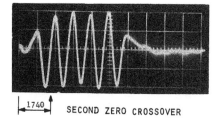

SECOND ZERO CROSSOVER

TRIGGER DELAY TIME: 19,000 MICROSECONDS
TOTAL TIME DELAY: 20,740 MICROSECONDS

FIGURE 5.21.   SECOND ZERO CROSSOVER OF WWV
              TICK AT A SWEEP RATE OF
              1 MILLISECOND/DIVISION.

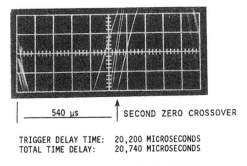

540 µs   SECOND ZERO CROSSOVER

TRIGGER DELAY TIME: 20,200 MICROSECONDS
TOTAL TIME DELAY: 20,740 MICROSECONDS

FIGURE 5.22.   SECOND ZERO CROSSOVER OF WWV
              TICK AT A SWEEP RATE OF 100
              MICROSECONDS/DIVISION (4 SWEEPS)

91

the nearest one-tenth of a millisecond and added to the trigger delay, resulting in an approximate total time delay. If the local clock 1 pps time is exactly coincident with the UTC(NBS) seconds pulse, the total measured time delay will be approximately equal to the sum of the propagation delay time, the receiver delay time (typically 200 - 500 microseconds), and the cycle correction (1000 microseconds for WWV or CHU and 833 microseconds for WWVH).

To further increase the resolution of delay measurement, the oscilloscope sweep rate can be increased to 0.1 ms/division (100 microseconds/division) and the trigger pulse from the generator adjusted to be approximately 500 microseconds less than the total delay time previously measured. At these settings, the second zero crossover of the tick will be somewhere near the midscale of the oscilloscope face.

The vertical centering of the sweep should be rechecked and centered if necessary. The tick is measured to the nearest 10 microseconds (figure 5.21). The result should be within ± 100 microseconds of the result obtained at the 1 ms/division sweep rate. If the result of this measurement falls outside this tolerance, then the procedure should be repeated again by measuring the total time delay at a sweep rate of 1 ms/ division. To obtain the time, the equations described earlier should be used.

E. Using Oscilloscope Photography for Greater Measurement Accuracy

By film recording five or more overlapping exposures of the WWV/WWVH tick, an average of the tick arrival time can be estimated

with more accuracy. The exposures are made when consistently strong and undistorted ticks appear on the oscilloscope. To determine the time, the average of the second zero crossover point of the tick is measured using the same procedure explained above.

In making measurements using this technique, an oscilloscope camera using self-developing film is mandatory. The camera shutter is placed in the time exposure position so that it can be opened and closed manually. The lens opening of the camera, the oscilloscope trace intensity, and the scale illumination must be determined by experiment. Refer to your camera manual.

One of the previously described procedures is followed to obtain the seconds tick. At a sweep rate of 1 ms/division, the shutter is opened before the sweep starts and closed after the sweep ends. This is repeated each second until five overlapping exposures are completed (figure 5.23). The pictures should be taken when the ticks begin to arrive with the least distortion and maximum amplitude.

This procedure can also be used at a faster sweep rate of 100 microseconds/division with the second zero crossover point appearing approximately at midpoint of the trace. (One complete cycle of the tick should be visible-- figure 5.24.) Overlapping exposures of the ticks .are taken and an average reading is obtained from the photograph.

5.4.3 USING THE WWV/WWVH TIME CODE

Among the several tones, ticks, and voice signals offered on WWV and WWVH, there is a time code. If you receive and decode this signal, you can automatically display the day

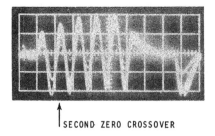

SECOND ZERO CROSSOVER

FIGURE 5.23.   WWV TICK AT A SWEEP RATE OF 1 MILLISECOND/DIVISION (5 OVERLAPPING EXPOSURES).

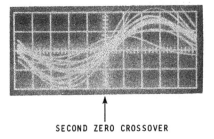

SECOND ZERO CROSSOVER

FIGURE 5.24.   WWV TICK AT A SWEEP RATE OF 100 MICROSECONDS/DIVISION (OVERLAPPING EXPOSURES)

of the year, hour, minute, and second. Of course, it is not nearly that easy--there is a lot more to it than simply receiving the code. To avoid errors, you must design your decoder/clock carefully.

Where is the code hidden? You will find it very near the carrier signal--only 100 Hz away, in fact. This is shown in figure 5.25. Also shown are other "parts" of the WWV spectrum. Although 10 MHz is used as an example, the same distribution applies to the other frequencies if you simply change the carrier.

This signal, 100 Hz away from the main carrier, is called a subcarrier. The code pulses are sent out once per second. With a good signal from a fairly high-quality receiver, you can hear the time code as a low rumble in the audio. It can also sometimes be heard on the NBS telephone time-of-day circuits.

A. Code Format

Now that we know where the pulses are and how they are sent, what do they say? First of all, they follow a specific format. In this case, it is a modified IRIG-H format similar to that shown in figure 1.2 of Chapter 1. After suitable identifiers are sent, the bits that make up the units--tens and hundreds for minutes, hours and days--are sent sequentially.

Certain pulses in succession comprise binary-coded groups representing decimal numbers. The binary-to-decimal weighting scheme is 1-2-4-8 with the least significant binary digit always transmitted first. The binary groups and their basic decimal equivalents are shown in the following table:

BINARY GROUP WEIGHT: 1 2 4 8	DECIMAL EQUIVALENT
0 0 0 0	0
1 0 0 0	1
0 1 0 0	2
1 1 0 0	3
0 0 1 0	4
1 0 1 0	5
0 1 1 0	6
1 1 1 0	7
0 0 0 1	8
1 0 0 1	9

In every case, the decimal equivalent of a BCD group is derived by multiplying each binary digit times the weight factor of its respective column and then adding the four products together. For instance, the binary sequence 1010 in the 1-2-4-8 scheme means (1×1) + (0×2) + (1×4) + (0×8) = 1 + 0 + 4 + 0 = 5, as shown in the table. If fewer than nine decimal digits are needed, one or more of the binary columns may be omitted.

In the standard IRIG-H code, a binary 0 pulse consists of exactly 20 cycles of 100-Hz amplitude modulation (200 milliseconds duration), whereas a binary 1 consists of 50 cycles of 100 Hz (500 milliseconds duration). In the WWV/WWVH broadcast format, however, all tones are suppressed for 40 ms while the seconds pulses are transmitted.

Because the tone suppression applies also to the 100-Hz subcarrier frequency, it has the effect of deleting the first 30-millisecond portion of each binary pulse in the time code. Thus, a binary 0 contains only 17 cycles of 100-Hz amplitude modulation (170 milliseconds duration) and a binary 1 contains 47 cycles of

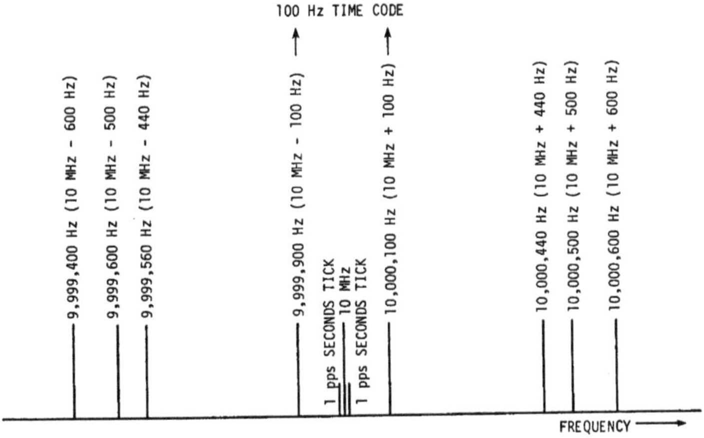

100 Hz TIME CODE

FIGURE 5.25. THE WWV SPECTRUM AT 10 MHz.

93

100 Hz (470 milliseconds duration). The leading edge of every pulse coincides with a positive-going zero crossing of the 100-Hz subcarrier, but it occurs 30 milliseconds after the beginning of the second.

Within a time frame of one minute, enough pulses are transmitted to convey in BCD language the current minute, hour, and day of year. Two BCD groups are needed to express the hour (00 through 23); and three groups are needed to express the day of year (001 through 366). When representing units, tens, or hundreds, the basic 1-2-4-8 weights are simply multiplied by 1, 10, or 100 as appropriate. The coded information always refers to time at the beginning of the one-minute frame. Seconds may be determined by counting pulses within the frame.

Each frame commences with a unique spacing of pulses to mark the beginning of a new minute. No pulse is transmitted during the first second of the minute. Instead, a one-second space or hole occurs in the pulse train at that time. Because all pulses in the time code are 30 milliseconds late with respect to UTC, each minute actually begins 1030 milliseconds (or 1.03 seconds) prior to the leading edge of the first pulse in the new frame.

For synchronization purposes, every ten seconds a so-called position identifier pulse is transmitted. Unlike the BCD data pulses, the position identifiers consist of 77 cycles of 100 Hz (770 milliseconds duration).

UT1 corrections to the nearest 0.1 second are broadcast via BCD pulses during the final ten seconds of each frame. The coded pulses which occur between the 50th and 59th seconds of each frame are called control functions. Control function #1, which occurs at 50 seconds, tells whether the UT1 correction is negative or positive. If control function #1 is a binary 0, the correction is negative; if it is a binary 1, the correction is positive. Control functions #7, #8, and #9, which occur respectively at 56, 57, and 58 seconds, specify the amount of UT1 correction. Because the UT1 corrections are expressed in tenths of a second, the basic binary-to-decimal weights are multiplied by 0.1 when applied to these control functions.

Control function #6, which occurs at 55 seconds, is programmed as a binary 1 throughout those weeks when Daylight Saving Time is in effect and as a binary 0 when Standard Time is in effect. The setting of this function is changed at 0000 UTC on the date of change. Throughout the U. S. mainland, this schedule allows several hours for the function to be received before the change becomes effective locally--i.e., at 2:00 am local time. Thus, control function #6 allows clocks or digital recorders operating on local time to be programmed to make an automatic one-hour adjustment in changing from Daylight Saving Time to Standard Time and vice versa.

Figure 5.26 shows one frame of the time code as it might appear after being rectified, filtered, and recorded. In this example, the leading edge of each pulse is considered to be the positive-going excursion. The pulse train in the figure is annotated to show the characteristic features of the time code format.

The six position identifiers are denoted by symbols P_1, P_2, P_3, P_4, P_5, and P_0. The minutes, hours, days, and UT1 sets are marked by brackets, and the applicable weighting factors are printed beneath the coded pulses in each BCD group. With the exception of the position identifiers, all uncoded pulses are set permanently to binary 0.

The first ten seconds of every frame always include the 1.03-second hole followed by eight uncoded pulses and the position identifier P_1. The minutes set follows P_1 and consists of two BCD groups separated by an uncoded pulse. Similarly, the hours set follows P_2. The days set follows P_3 and extends for two pulses beyond P_4 to allow enough elements to represent three decimal digits. The UT1 set follows P_5, and the last pulse in the frame is always P_0.

In figure 5.26, the least significant digit of the minutes set is $(0 \times 1) + (0 \times 2) + (0 \times 4) + (0 \times 8) = 0$; the most significant digit of that set is $(1 \times 10) + (0 \times 20) + (0 \times 40) = 10$. Hence, at the beginning of the 1.03-second hole in that frame, the time was exactly 10 minutes past the hour. By decoding the hours set and the days set, it is seen that the time of day is in the 21st hour on the 173rd day of the year. The UT1 correction is +0.3 second. Therefore, at point A, the correct time on the UT1 scale is 173 days, 21 hours, 10 minutes, 0.3 second.

B. Recovering the Code

The time code will appear at the detector in any receiver tuned to WWV or WWVH. However, due to the limited low frequency passband of most receivers, it may or may not be heard at the speaker or on headphones. Receivers are deliberately designed to suppress lower frequencies to limit the power supply hum (60 Hz) and its second harmonic (120 Hz). The best approach would be to start at the receiver-detector and "strip" out the 100 Hz with a narrow filter. If the 100 Hz is then amplified and detected, a code stream can be produced and used to drive a decoder/display.

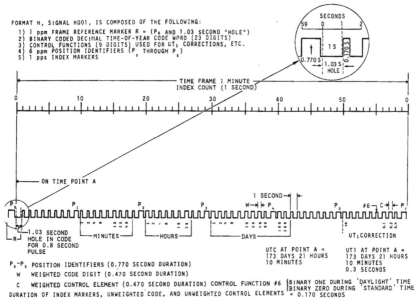

FORMAT H, SIGNAL H001, IS COMPOSED OF THE FOLLOWING:

1) 1 ppm FRAME REFERENCE MARKER R = (P₀ AND 1.03 SECOND "HOLE")
2) BINARY CODED DECIMAL TIME-OF-YEAR CODE WORD (23 DIGITS)
3) CONTROL FUNCTIONS (9 DIGITS) USED FOR UT₁ CORRECTIONS, ETC.
4) 6 ppm POSITION IDENTIFIERS (P₀ THROUGH P₅)
5) 1 pps INDEX MARKERS

P₀–P₅ POSITION IDENTIFIERS (0.770 SECOND DURATION)
W   WEIGHTED CODE DIGIT (0.470 SECOND DURATION)
C   WEIGHTED CONTROL ELEMENT (0.470 SECOND DURATION) CONTROL FUNCTION #6 {BINARY ONE DURING 'DAYLIGHT' TIME
DURATION OF INDEX MARKERS, UNWEIGHTED CODE, AND UNWEIGHTED CONTROL ELEMENTS = 0.170 SECONDS   {BINARY ZERO DURING 'STANDARD' TIME

NOTE: BEGINNING OF PULSE IS REPRESENTED BY POSITIVE-GOING EDGE.

9/75

FIGURE 5.26. WWV/WWVH TIME CODE FORMAT.

WWV decoders can be purchased or home-built. Some hobby and radio amateur publications have carried articles on using the WWV time code. At this writing, the NBS is not able to supply decoder designs. As such material becomes available, it will be announced in the NBS Time and Frequency Bulletin, published monthly.

Because signals such as those transmitted by WWV tend to fade, it is possible to lose some of the code bits. As these errors occur, any clock driven directly from a WWV receiver would display incorrect time. This can be overcome by having the decoder circuit detect the errors. One scheme for error detection simply decodes one frame of the incoming code and stores it electronically as digital bits. The next frame is received and stored in another location. If the two frames do not differ by exactly one minute, an error has occurred. The first data are discarded and another try is made until several successful decodes have been made. This can be carried out to three, four, or five successive, suc-cessful decodes to reduce the possibility of error.

Meanwhile, the clock can be made to operate from a simple crystal oscillator. It will operate from that crystal until the time code has been successfully decoded and is used to reset the clock to the correct time if necessary. When you consider how accurate crystal clocks are, the code really doesn't have very much to do, but it will reset the clock as needed and automatically adjust for Daylight Saving Time. It will also increase your confidence in the clock as a timekeeper. There are 60 minutes and therefore 60 frames in the WWV time code every hour. This means that there are 1440 frames in a day. A number of successful decodes should be received in a day. Typically, this means that the crystal clock would be corrected several times. This should be more than adequate for good time-keeping.

An electronic means can be provided to receive and decode Coordinated Universal Time but display local time as needed.

95

5.5 USE OF HF BROADCASTS FOR
FREQUENCY CALIBRATIONS

In addition to the widely used time service of WWV and WWVH, standard frequencies are also available to the broadcast listener. With a general purpose HF receiver capable of tuning these stations, a calibrating frequency is readily available for comparison and measurement.

Frequency or its inverse, time interval, is an important quantity in physics, electronics, and in our everyday lives. From tuning musical instruments to setting the radio and TV dials, frequency enters our lives in many ways. If we say we are going to perform a frequency calibration, the question arises, on what? If we calibrate the frequency source in our watch, we have to adjust the balance wheel (or the crystal). Radio listeners often want to check the dial settings on their shortwave receivers. That way, they know if there is an error in the dial when they are seeking a certain station whose transmitter frequency is known.

There are two ways to use a frequency source such as WWV. On the one hand, you can measure your frequency source against WWV and write the difference down for reference, or you can actually readjust the source of frequency you have to make it agree with WWV.

WWV and WWVH broadcast low frequency tones at 440, 500, and 600 Hz. These can be used with an oscilloscope or electronic counter for checking audio oscillators, musical instruments, and other devices that operate at low frequencies. The accuracies obtainable are usually satisfactory for these applications but would not be sufficient for setting a crystal oscillator.

In the material that follows, several calibration methods are discussed. Generally, WWV-type signals are used to calibrate lower accuracy frequency sources with 1 part in 10^7 being about the limit under good conditions. Also, the HF radio services lend themselves well to the user who is willing to turn the knob and change his oscillator frequency. These methods are best used to set two sources equal rather than to measure the numerical difference between two frequency sources.

Direct frequency comparison or measurement with WWV can usually be accomplished to about one part in 10^6. This resolution can be improved by long-term (many weeks) time comparison of clocks operated from a frequency source rather than direct frequency comparison. Four methods of calibrating frequency sources using the broadcasts of WWV/WWVH are discussed: (1) beat frequency method; (2) oscilloscope Lissajous pattern method; (3) oscilloscope pattern drift method; and (4) frequency calibrations by time comparisons.

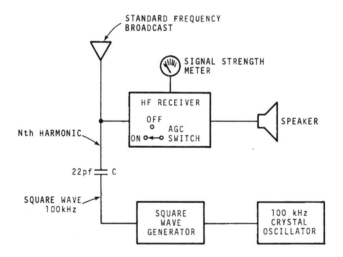

FIGURE 5.27. EQUIPMENT SETUP FOR BEAT FREQUENCY METHOD OF CALIBRATION.

96

5.5.1 BEAT FREQUENCY METHOD

The beat frequency or heterodyne method of frequency comparison with standard radio frequencies is a simple technique commonly used by radio operators to calibrate transmitters and tune receivers. A frequency offset of about 1 part in 10^6 can be accurately determined. Thus, a 1 MHz signal that is calibrated in this way can have an expected error of 1 Hz.

Figure 5.27 shows an arrangement for calibrating a 100 kHz oscillator. A 100 kHz signal containing harmonics is coupled to the receiver input along with the WWV signal from the antenna.

This method works by heterodyning or mixing a known and accurate frequency (e.g., WWV RF signal) with the output of an oscillator. The mixing is accomplished by the converter circuit in any superheterodyne receiver. The difference frequency of the two RF signals can be amplified and detected. The result is an audio output signal called the beat frequency or beat note.

The frequency of this beat note is the difference of the two input frequencies. When the two frequencies are made equal, their difference decreases to zero and is called "zero beat." When zero beat is reached, the oscillator is equal in frequency to the WWV frequency.

To calibrate a frequency source or crystal oscillator with an output frequency lower than that broadcast by WWV, the correct harmonic equal to the WWV signal is required. For example, if a 100 kHz signal is to be calibrated with the WWV 5 MHz carrier frequency, then it must also contain a harmonic fifty times itself. Thus, a signal to be calibrated must be a submultiple of the WWV carrier frequency.

Theoretically, a sine wave does not contain any harmonics. In practice, though, all sine wave signals contain sufficient harmonics to produce a beat note. A square wave signal, on the other hand, is very rich in harmonic content and is ideal for generating harmonics to calibrate receivers and transmitters in the HF and VHF band.

A simple method of generating a square wave from a sine wave is by clipping the signal with a diode clipping circuit--shown in figure 5.28. To obtain a strong harmonic signal for beat notes requires a large amplitude signal to produce heavy clipping. A better method is to digitally condition the 100 kHz signal to produce square waves.

If the receiver input impedance is low (50 -100 ohms), a 10 to 20 picofarad (pf) capacitor can be used to couple the high frequency harmonic to the receiver input and to reduce the level of the lower fundamental frequency. If the receiver has a high input impedance with unshielded lead-in wire from the antenna, the harmonic signal can be loosely coupled to the receiver input by wrapping a few turns of an insulated wire around the antenna lead-in and connecting it directly to the output of the oscillator. For receivers with built-in or whip antennas, you must experiment to find a way to inject the oscillator signal.

Using harmonics of the oscillator being calibrated makes it necessary to know the relationship between the oscillator error and the beat note that is measured during calibration. Let the oscillator output be designated as f_F. This is made up of two components, the correct frequency f_0, plus an error that we can designate as Δf. So:

$$f_F = f_0 + \Delta f.$$

A harmonic of this oscillator signal (written Nf_F) will beat against the carrier, f_C. The resulting beat note f_B is the difference between the two, written as:

$$f_B = f_C - Nf_F .$$

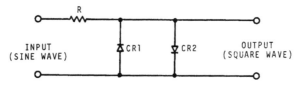

CR1, CR2 = 1N270, 1N34

FIGURE 5.28. DIODE CLIPPING CIRCUIT TO PRODUCE HARMONICS FROM A PURE SINE WAVE SIGNAL.

The bars mean that a negative sign in the answer is ignored; that is, the answer is always a positive number.

Now substitute the first equation into the second:

$$f_B = f_C - Nf_0 - N\Delta f_F .$$

But Nf_0 equals f_C. That's why we multiplied by N. So:

$$f_B = N\Delta f, \text{ or}$$

$$\Delta f = \frac{f_B}{N} ,$$

which says that the error equals the beat that we hear divided by N. For example, if a beat frequency of 100 Hz were measured between the WWV 5 MHz signal and the 50th harmonic of a 100 kHz oscillator signal, the frequency error of the 100 kHz signal would be:

$$\Delta f = \frac{100 \text{ Hz}}{50} = 2 \text{ Hz.}$$

The oscillator frequency is in error by 2 Hz. To determine whether the oscillator is high or low in frequency, the oscillator frequency has to be changed to note which way the beat frequency decreases. If increasing the oscillator frequency decreases the beat note, it indicates that the oscillator frequency is lower than the WWV/WWVH frequency.

In a receiver with no tuned RF amplifier between the mixer stage and the antenna input, a low frequency sine wave signal can enter the mixer stage directly and generate unwanted harmonics and confusing beat notes due to the nonlinear characteristics of a mixer circuit. However, a good communications receiver or a special-purpose WWV receiver generally has tuned RF amplifier stages or preselectors before its first mixer stage. Then, only the desired beat note from two input signals is produced.

If the beat note is above 50 Hz, head-phones, a speaker, or a counter can be used. Below that frequency, a dc oscilloscope can be connected to the receiver detector. A signal strength meter can be used and the beats counted visually. The automatic gain control (AGC) should be disabled, if possible, for the meter fluctuations to be more noticeable. The manual RF gain can be adjusted to compensate for loss of the AGC.

To correct the oscillator frequency, the adjustment knob is turned in the direction

which lowers the frequency of the beat note. Between 50 Hz and about 1 Hz, the beat note cannot be heard and the signal strength meter will begin to respond to the beat note as it approaches 1 Hz. As zero beat is approached, a very slow rise and fall of the background noise or the WWV audio tone can also be heard on the speaker. The meter effect is much easier to follow. As it approaches zero beat, the very slow rise and fall of the signal strength may sometimes become difficult to distinguish from signal fading due to propagation effects.

To overcome fading effects, the oscilla-tor adjustment can be interpolated. First, adjust the oscillator to a minimum beat fre-quency that can be measured without interfer-ence. For accuracy, count the number of de-flections of the signal strength meter in 10 seconds. The setting of the frequency adjust-ment is then marked. The adjustment is then made to pass zero beat until the beat is again visible on the meter. By obtaining the same number of meter deflections as the previous beat note, the frequency can be set midway between the two adjustments.

The frequency of crystal oscillators changes with time. This is commonly referred to as the aging or drift rate. Therefore, all crystal oscillators must be recalibrated peri-odically if maximum accuracy is desired.

5.5.2 OSCILLOSCOPE LISSAJOUS PATTERN METHOD

Audio oscillators can be calibrated by using WWV/WWVH audio tones to produce Lissa-jous patterns on an oscilloscope. Lissajous patterns are fully discussed in Chapter 3. The WWV audio signal is applied to the verti-cal input of the scope and the oscillator signal to be calibrated is used to drive the horizontal amplifier. The resultant pattern tells the user two things: (1) the frequency ratio between his oscillator setting and the received tone, and (2) movement of the oscil-lator relative to WWV.

The tones available on WWV nd WWVH are mainly 500 and 600 Hz. CHU transmits 1 kHz ticks that last for about one-third second. If these are used, they will produce an oscil-loscope pattern at the tick rate; i.e., once per second. Even the WWV/WWVH ticks can be used this way. Most important, the user should be aware of the exact format being broadcast to avoid making errors. This can be a problem because WWV and WWVH switch between 500 and 600 Hz in alternate minutes and the 440 Hz is used for only one minute every hour.

In a typical application, the user will be able to check the accuracy of the dial operation. First, he picks a dial setting giving a frequency ratio to a WWV tone that is an integer. Then he turns the dial slowly until the pattern is stationary. By reading the dial setting, a calibration can be made and the dial reset to another frequency that is an integer ratio. The same technique could be applied to fixed frequency sources if they are in correct ratio to the tones on WWV and WWVH. The ratio of the two frequencies is equal to the ratio of the number of loops along two adjacent edges. If the vertical input frequency is known, the horizontal input frequency can be expressed by the equation:

$$f_h = \frac{N_v}{N_h} f_v \, ,$$

Where: f_h = horizontal input frequency

f_v = vertical input frequency

N_v = number of loops on the vertical edge

N_h = number of loops on the horizontal edge

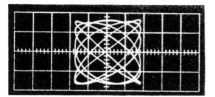

VERTICAL INPUT: WWVH 500 Hz TONE
HORIZONTAL INPUT: AUDIO OSCILLATOR
                   600 Hz TONE

FIGURE 5.29. OSCILLOSCOPE DISPLAY SHOWING LISSAJOUS PATTERN.

The pattern shown in figure 5.29 has five loops on the horizontal edge and six loops on the vertical edge. The vertical input frequency is the WWVH 500 Hz tone. To determine the horizontal input frequency, the known values are substituted in the above equation and the result is:

$$f_h = \frac{6}{5} \times 500 = 600 \text{ Hz.}$$

Therefore, the horizontal input frequency is 600 Hz.

It is possible to calibrate over a ten-to-one range in frequency both upwards and downwards from the 500 and 600 Hz audio tones transmitted by WWV and WWVH; that is, from 50 Hz to 6 kHz. However, not all frequencies between them can be calibrated with the 500 and 600 Hz tones because not all frequencies have a ratio of integers with less than the number ten in the numerator and denominator. For example, a frequency of 130 Hz compared with the 500 Hz tone would give a ratio of 50:13. It would be impossible to count 50 loops on the horizontal edge. But a frequency ratio of 500 Hz to 125 Hz is possible because there will be four loops on the horizontal edge and only one loop on the vertical edge or a ratio of 4:1. A frequency ratio of 1:1 produces a less complex pattern of a tilted line, circle, or ellipse.

If the frequencies are exactly equal, the figure will remain stationary. If one frequency is different from the other, the figure will not remain stationary but will "rotate." Because one complete "rotation" of the figure is equal to one cycle, the number of cycles per unit of time is the frequency error.

For example: If a Lissajous figure takes ten seconds to "rotate" through one cycle and the frequency being compared is 600 Hz, the frequency error is:

$$\Delta f = \frac{1}{T} = \frac{1}{10} = 0.1 \text{ Hz}$$

$$F = \frac{\text{Relative}}{\text{Frequency}} = \frac{\Delta f}{f} = \frac{0.1}{600} = 1.7 \times 10^{-4}$$

Since the error is inversely proportional to the time it takes to complete one cycle, it is obvious that the longer it takes to complete a cycle, the smaller the error will be.

The measurement accuracy of a Lissajous pattern by the comparison method described above is inversely proportional to the frequency. For example, if the time required for a Lissajous figure to "rotate" through one cycle is ten seconds at a frequency of 1 MHz:

$$F = \text{Relative Frequency} = \frac{\Delta f}{f}$$

$$= \frac{0.1 \text{ Hz}}{1 \times 10^6 \text{ Hz}} = 1 \times 10^{-7}$$

If, instead of measuring at 1 MHz, we observe the Lissajous pattern at 1 kHz:

$$\Delta f = f \times F = 1 \times 10^3 \times 1 \times 10^{-7} = 10^{-4} \text{ Hz}$$

99

and the observation time would be:

$$\tau = \frac{1}{\Delta f} = \frac{1}{10^{-4}} = 10,000 \text{ seconds.}$$

Thus, it would take too long to measure signals with relative frequencies of less than 1×10^{-5} at 1 kHz. A more accurate method which measures the phase shift on an oscilloscope will be discussed next.

5.5.3 OSCILLOSCOPE PATTERN DRIFT METHOD

The oscilloscope pattern drift method is good for comparing two frequencies if you have an oscilloscope with external triggering. It can detect smaller frequency differences than the Lissajous pattern at audio frequencies.

The method consists of an oscilloscope with an accurately calibrated sweep time base. External triggering is obtained from the signal to be calibrated. This signal can be any integer submultiple of the tone being received from WWV/WWVH. The receiver (tuned to WWV) has its audio output connected to the vertical input of the oscilloscope. With the sweep rate set at 1 millisecond/division, the trigger level is adjusted so that a zero crossover of the corresponding 600 or 500 Hz signal is near midscale on the scope.

By measuring the phase drift during a given time interval, the frequency error is determined. If the zero crossover moves to the right, the audio signal frequency is higher than the WWV signal, and if it moves to the left, the signal is lower in frequency (fig. 5.30).

For example, if during a count of 10 seconds at 500 Hz, the zero crossover advanced from left to right by 0.1 millisecond, the relative frequency is:

$$F = \frac{\text{phase drift}}{\text{time interval}} = \frac{+0.1 \times 10^{-3}}{10} = +1 \times 10^{-5},$$

and the frequency difference is

$$\Delta f = f \times 1 \times 10^{-5} = 500 \times 10^{-5} = +0.005 \text{ Hz.}$$

Therefore, the 500 Hz signal has an error of +0.005 Hz. Under ideal conditions, a relative frequency of 1 part in 10^6 can be determined by increasing the observation time.

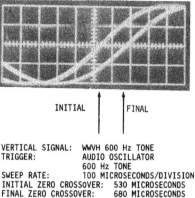

INITIAL   FINAL

VERTICAL SIGNAL:	WWVH 600 Hz TONE
TRIGGER:	AUDIO OSCILLATOR 600 Hz TONE
SWEEP RATE:	100 MICROSECONDS/DIVISION
INITIAL ZERO CROSSOVER:	530 MICROSECONDS
FINAL ZERO CROSSOVER:	680 MICROSECONDS
PERIOD OF MEASUREMENT:	10 SECONDS
TOTAL PHASE SHIFT:	+150 MICROSECONDS

$$F = \frac{+150 \times 10^{-6}}{10} = +1.5 \times 10^{-5}$$

$$\Delta f = 600 \times 1.5 \times 10^{-5} = 900 \times 10^{-5} = +0.009 \text{ Hz}$$

FIGURE 5.30. OSCILLOSCOPE DISPLAY USING PATTERN DRIFT METHOD.

5.5.4 FREQUENCY CALIBRATIONS BY TIME COMPARISON OF CLOCKS

For those users who are set up to recover and maintain time from WWV or WWVH, there is an alternate way to calibrate the frequency of the oscillator driving the clock. This is simply to record the time differences from day to day and then calculate the frequency rate. The operation is analogous to telling your jeweler that your watch gains or loses so many seconds a day. He then adjusts the rate to compensate.

The operation proceeds something like this. Each day, note the amount the clock output differs from WWV. Then keep an accurate record of the history of the oscillator in terms of the time gained or lost with respect to WWV ticks. To get accurate data, readings from several days should be averaged.

For example: If the time error averaged out to be 1 millisecond per day (say for a 5-day period), then you could compute the frequency offset by noting that (see Chapter 3):

$$\frac{\Delta f}{f} = - \frac{\Delta t}{T}$$

100

which simply says that the time lost or gained by a clock whose oscillator is running at frequency f is proportional to the frequency error. In our example, putting all the terms in the same units (in this case, seconds) we get:

$$\frac{\Delta t}{T} = \frac{5 \times 10^{-3} \text{ s time error}}{5 \times 86,400 \text{ s}}$$

$$= 1.2 \times 10^{-8} = -\frac{\Delta f}{f}$$

F, the relative frequency of this oscillator would be -1.2×10^{-8}. So an oscillator that is off frequency by about a part in 10^8 would gain or lose about a millisecond a day. Using the techniques outlined in the section on time checking, this millisecond error would be quite visible to a user.

This example illustrates that this technique could be extended to other levels of accuracy. In fact, data have been obtained that allow WWV signals to be used for calibrations approaching a part in 10^{10}.

We can use our example to illustrate another side of the time/frequency problem; namely, how to specify frequency error. Assuming that our answer was one part in 10^8, we can write it mathematically as 1×10^{-8}, and (if the oscillator were 100 kHz and high in frequency) our actual numerical frequency would be 100,000.001 hertz. This can also be written as 0.000001% frequency error.

In summary then, frequency calibrations can be made by recording only the time error produced by an oscillator. This technique is recommended for lower quality oscillators and for situations where the user is already measuring time of day from WWV for some other reason.

5.6 FINDING THE PROPAGATION PATH DELAY

The problem of finding the radio propagation path delay is twofold. First, the great circle (curved) distance must be found between the transmitter and receiver. Second, the number of times the signal bounces from ionosphere to earth (hops) must be known. With these two facts, we can solve for the total path delay. The accuracy we can expect will be somewhere around a millisecond. We would be hard put to improve that figure very much. For groundwave signals at very low frequencies, we could carry a clock and calibrate the

path, but for HF waves of the type transmitted by WWV and CHU, it probably isn't worth the effort. The reason for this is that at any particular time, the path may not be at its average value. Since the path is constantly changing by small amounts, it would not pay to try to measure it with great exactness.

For those users who want time with the best possible accuracy and who need a method for calculating the path delay, the following methods will work. The new hand-held calculators that are available will greatly assist in the computations. However, the math can be done with a slide rule or a book of tables. Some examples are given to help in the solution..

5.6.1 GREAT CIRCLE DISTANCE CALCULATIONS

Using figure 5.31 as a reference:

A and B are two points on the earth.

Pole is one of the two poles, North or South.

L_A = Latitude at point A.

L_B = Latitude at point B.

L_{OA} = Longitude at point A.

L_{OB} = Longitude at point B.

P = $L_{OA} - L_{OB}$.

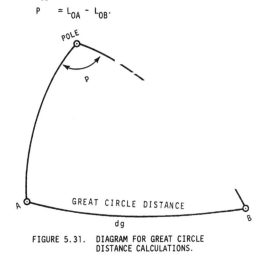

FIGURE 5.31. DIAGRAM FOR GREAT CIRCLE DISTANCE CALCULATIONS.

PROBLEM: Using the above information, find the great circle distance (dg) in nautical miles between the two points A and B. All angles are given in degrees.

For points A and B on the same side of the equator:

$$dg = 60 \cos^{-1} (\cos L_A \cos L_B \cos P$$
$$+ \sin L_A \sin L_B) \text{ nautical miles.}$$

For points A and B on opposite sides of the equator:

$$dg = 60 \cos^{-1} (\cos L_A \cos L_B \cos P$$
$$- \sin L_A \sin L_B) \text{ nautical miles.}$$

CONVERSION TO STATUTE MILES AND KILOMETERS:

1.151 x nautical miles = statute miles.
1.8522 x nautical miles = kilometers.

EXAMPLE: Find the great circle distance from radio station WWVH, Kauai, Hawaii (point A) to WWV, Fort Collins, Colorado (point B) in statute miles and kilometers.

The coordinates of the two stations are:

Latitude	Longitude
$L_A = 21°59'26'' \text{ N}$	$L_{OA} = 159°46'00'' \text{ W}$
$L_B = 40°40'49'' \text{ N}$	$L_{OB} = 105°02'27'' \text{ W}$

Then, $P = 159°46'00'' - 105°02'27'' = 54°43'33''$

SOLUTION: Taking sines and cosines:

$\sin L_A = 0.37445$ $\cos L_A = 0.92725$
$\sin L_B = 0.65184$ $\cos L_B = 0.75836$
$\cos P = 0.57749$

$$dg = 60 \cos^{-1} [(0.92725 \times 0.75836 \times 0.57749)$$
$$+ (0.37445 \times 0.65184)]$$

$$= 60 \cos^{-1} [(0.40608) + (0.24408)]$$

$$= 60 \cos^{-1} 0.65017 = 60 \times 49.44586$$

$$= 2966.75 \text{ nautical miles}$$

$$= 3414.73 \text{ statute miles}$$

$$= 5495.00 \text{ kilometers}$$

5.6.2 PROPAGATION DELAYS

The delay of high frequency radio waves depends on many factors. The height of the ionosphere and the distance between transmitter and receiver are important. Knowing the exact number of hops is important; otherwise, errors of from 0.5 to 1 millisecond can occur.

The ionosphere is made up of a number of layers whose actual height varies both daily and seasonally. This changes the path delay a lot if the receiver and transmitter are close to each other. Much of the time, such changes are not important to the person who wants to get time information, but if you are concerned with errors of about 300 microseconds, you must take care to account for all delays.

Since the earth is round, there is a maximum ground distance a single hop can span (fig. 5.33). This is about 4000 kilometers or 2200 miles. For greater distances, the radio waves must obviously be reflected a number of times. Any error in the wave path computation for one hop must be multiplied by the total number of hops.

For minimum error, the frequency used for reception should be selected for the least number of hops. This also results in a stronger signal due to smaller losses. At any particular distance, the user must select the proper shortwave frequency from WWV/WWVH so the signal can be received. Some frequencies will penetrate the ionosphere and not reflect. Other frequencies will skip over the user. He is then said to be located in the "skip distance" for that frequency. A lower frequency must be selected. Use of the maximum frequency that is receivable assures the least number of hops. This maximum usable frequency is called the MUF. A higher frequency will skip over the receiving station. A frequency about 10% below the MUF will provide the best reception.

For distances under one thousand miles, the single-hop mode of transmission dominates. For short distances, the height must be

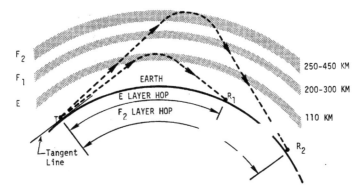

FIGURE 5.32.   REFLECTION OF RADIO WAVES AT DIFFERENT IONIZED LAYERS.

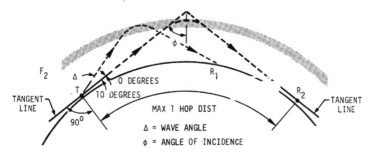

FIGURE 5.33.   SINGLE-HOP REFLECTIONS FROM F_2 LAYER AT DIFFERENT WAVE ANGLES.

estimated carefully. For distances of over 1600 kilometers (1000 miles), where only a single-hop mode of transmission occurs, a much wider range of error in the height can be tolerated.

By estimating height according to time of day, season, latitude, and sunspot cycle, errors can be reduced. For F_2-layer reflections, which is the principal layer used, there are several "averages" in height that can be used. An average height of 250 km is reasonable for the winter months when propagation conditions are good at 10 MHz and above. In summer, the average height can be increased to 350 km. As reception again begins to improve in the fall, the estimated height can be decreased again. Using this approximation method, the error in path length computation can be reduced considerably.

For distances from one to two thousand miles (3220 km), where multiple-hop transmis-

sion occurs, the method above produces errors. For best results, the height must be determined with greater accuracy. Heights for a particular latitude, longitude, and time of day are available from the World Data Center, NOAA, Department of Commerce, Boulder, CO 80303.

By performing a single time comparison with a station, the probability of error in the hundreds of microseconds becomes very great due to the changes in propagation. A high degree of accuracy for a single measurement is remote. But, by conducting measurements at several times over a period of a week and by averaging with methods such as the moving average method, errors can be reduced.

The following equations for determining wave angle and propagation time delay have been simplified and are approximate, giving errors of 2% or less.

103

DEFINITIONS OF SYMBOLS

Δ = wave angle or angle of radiation in degrees.

ϕ = angle of incidence in degrees.

θ = 1/2 central angle subtended by radius R from A to B.

dg = great circle distance from transmitter to receiver in kilometers.

N = number of hops to cover distance dg.

$\frac{dg}{N}$ = great circle distance per hop (km/hop).

h' = virtual height of ionized layer (km).

R = 6,370 km, mean radius of earth.

d_p = total propagation distance (km).

TD_p = propagation time delay (milliseconds).

WAVE ANGLE EQUATIONS

$$\Delta = \tan^{-1} \left(\frac{h'}{R \sin \theta} + \tan \frac{\theta}{2} - \theta \right)$$

$$\theta(degrees) = \frac{dg}{2RN} \left(\frac{180}{3.14} \right) = 0.0045 \frac{dg}{N}$$

For $\frac{dg}{N}$ < 4000 km or θ < 18 degrees:

$$\Delta = \tan^{-1} \left(\frac{2Nh'}{dg} + \frac{dg}{4RN} \right) - 0.0045 \frac{dg}{N}$$

For θ < 20 degrees:

$$\Delta \cong \left(\frac{2Nh'}{dg} - \frac{dg}{4RN} \right) 57.3$$

PROPAGATION TIME DELAY

$$TD_p = \frac{2R \sin \theta}{299.8 \sin \phi} = \frac{dg}{299.8 \sin \phi}$$

(R and dg in km)

Where:

$$\phi = 90° - \tan^{-1} \left(\frac{h'}{R \sin \theta} + \tan \frac{\theta}{2} \right)$$

For $\frac{dg}{N}$ < 4000 km, θ < 18 degrees:

$$\phi = 90° - \tan^{-1} \left(\frac{2Nh'}{dg} + \frac{dg}{4RN} \right) \quad \text{(approximately)}$$

SAMPLE CALCULATIONS

EXAMPLE: . Given: dg = 3220 km

h' = 250 km

N = 2

Determine wave angle Δ and propagation time delay TD_p.

$$\Delta = \tan^{-1} \left(\frac{2Nh'}{dg} + \frac{dg}{4RN} \right) - 0.0045 \frac{dg}{N} = 13.25°,$$

Or, by the approximate method:

$$\Delta \cong \left(\frac{2Nh'}{dg} - \frac{dg}{4RN} \right) 57.3 = 14.2°$$

$$TD_p = \frac{dg}{299.8 \sin \phi}$$

But:

$$\phi = 90° - \tan^{-1} \left(\frac{2Nh'}{dg} + \frac{dg}{4RN} \right) = 69.5°$$

So:

$$TD_p = \frac{3220}{299.8 \sin 69.5} = 11.5 \text{ ms}$$

EXAMPLE: Given: dg = 3220 km

h' = 334 km

N = 1

The actual time you hear on the telephone will differ by a few milliseconds from that heard over WWV via radio. This is because of the delay in the telephone lines. In addition, if you are calling from any great distance, the way in which the telephone company handles your call limits the total accuracy of the WWV tone to a few hertz. This means that if you are trying to make a calibration at 1000 Hz, you could have an error as great as three cycles per second. This accuracy is not as good as that which can be achieved using radio methods. Users are warned about this limit of accuracy over the telephone.

Because the telephone service is very often fully utilized, you may get a busy signal. If this happens, wait a few minutes and redial. After your call is connected, the system will automatically hang up in three minutes unless you hang up first.

5.8 SUMMARY

As stated early in this chapter, HF radio broadcasts are commonly used for time and frequency calibrations. Accuracies ranging from 1 ms for time-of-day to 1 part in 10^7 for frequency can be obtained. The NBS telephone time-of-day service can also be used when accuracy requirements are not critical. However, if your requirements are higher, you should probably use one of the other methods discussed in the following chapters.

Along with the widely used high frequency radio transmissions of WWV and WWVH, the National Bureau of Standards operates a frequency and time service on WWVB, a low frequency (LF) radio station. WWVB operates at 60 kHz to take advantage of the stable radio paths in that frequency range. Both frequency and time signals are provided, but no voice transmissions are made due to the narrow bandwidth of the transmitter/antenna combination. Many countries have services in this band, which ranges from 30 to 300 kHz, as well as in the VLF (very low frequency) band from 3 to 30 kHz. In addition to discussing the use of WWVB, this chapter also covers other VLF stations throughout the world. The Loran-C service at 100 kHz is discussed in a Chapter 9.

It may seem unusual to send signals in a frequency band that is almost in the audio range, and indeed, these signals do pose some special problems for the transmitter and receiver-design engineers. However, low frequencies such as the 60 kHz of WWVB are used because of their remarkable stability. Radio waves at low frequencies use the earth and the ionosphere as a waveguide and follow the earth's curvature for long distances. Accuracies of a part in 10^{11} or better for frequency and 500 microseconds for time can be achieved by using LF or VLF broadcasts.

6.1 ANTENNAS FOR USE AT VLF-LF

At the transmitters, it is difficult to radiate high power levels. Special large antenna arrays are used, but even they are inefficient. Resonance of receiving antennas is hard to achieve because of the low frequency and resulting long wavelength. A channel 9 TV antenna is about 2 feet long. At 60 kHz, an antenna has to be almost 2 miles long (a quarter-wavelength) to be resonant. This is obviously impossible so a compromise must be made.

Since a quarter-wavelength antenna is out of the question, antennas that are electrically "short" are used with tuning boxes and special couplers. On vertical antennas, top loading is often used. This consists of radial wires extending from the top of the antenna to the ground. Often, they serve as guys. Of course, as with any antenna system, a good ground is essential. The physical location of antennas is important. Placement will affect signal strength and noise. Keep the antenna away from metal objects. Long-wire antennas should be at least 15 to 20 feet

above ground. Several commercial antennas use preamplifiers so the connecting coaxial cable supplies power to the amplifier. In those cases, care must be taken to avoid shorting the cable.

Manufacturers of VLF-LF radio receivers use a variety of antenna types. Long-wire antennas up to several hundred feet are available. Whip antennas 8 to 10 feet long are used where space is a problem. On the other hand, air loop antennas are able to reject interference but do not have as much gain as whips. This is done by turning the loop antenna so that the null of its figure 8 pattern is pointed at the source of interference. Ferrite loop antennas are becoming more popular since they are very small compared to an air loop, but they can cause some unwanted phase shift due to temperature. If you want the very best phase record, the kind of antenna and how it is used are important considerations.

All of the above antennas benefit from having couplers and/or amplifiers incorporated in the antenna structure to allow a match to be made with the shielded antenna cable. The usual precautions about placing antennas near metal objects or buildings should be observed. Here, the manufacturers' instructions should be followed. Many engineering handbooks and antenna manuals contain details about antenna construction.

6.2 SIGNAL FORMATS

Even after the antenna problems are solved, sending information on low-frequency carriers leaves much to be desired. WWVB sends a carrier for frequency information and changes the level (-10 decibels (dB)) of that carrier to transmit a time code in the form of binary "ones" and "zeros." The Navy VLF stations use on/off keying to send code and sometimes use frequency-shift or minimum-shift keying. The Omega Navigation System sends only the carrier. Receivers for these VLF signals usually "lock" onto the carrier and thus recover frequency information. In addition, some means is provided to hear the station. This is usually just a tone output. For example, if you were to listen to WWVB, you would hear the one-second code segments as a tone that changes from loud to soft. There are no voice signals on any of the LF or VLF stations. The bandwidth used cannot transmit voice signals. Almost all VLF-LF stations are controlled by atomic oscillators.

107

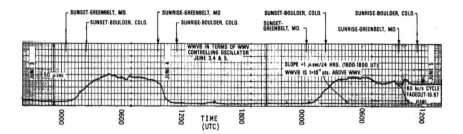

FIGURE 6.1. PHASE OF WWVB AS RECEIVED IN MARYLAND.

6.3 PROPAGATION CHARACTERISTICS AND OTHER PHASE CHANGES

Phase records made of VLF and LF stations show changes caused by the daily and seasonal changes in the propagation path. Among other things, LF and VLF signals have regular phase shifts. These shifts occur when sunrise or sunset occurs on the path from the transmitter to receiver. For instance, as the path is changing from all darkness to all daylight, the ionosphere lowers. This shortens the path from transmitter to receiver. This shortened path causes the received phase to advance. This phase advance continues until the entire path is in sunlight. The phase then stabilizes until either the transmitter or receiver enters darkness. When this happens, the ionosphere begins to rise, causing a phase retardation. Figure 6.1 shows a strip chart recording of the phase of WWVB, Ft. Collins, Colorado, as received in Maryland. The day/night changes are easily seen. The magnitude of the change is a function of the path length, and the rate of the change is a function of path direction.

A phase recording from a stable VLF or LF radio station contains a great deal of information. The user's job is to sort out this information so that he can understand what is happening to the frequency source that is being calibrated. His first assumption is usually that the signal from the station is almost perfectly controlled. That is, it starts on frequency and stays there. This is not always true. Errors in control at the transmitter can cause large phase shifts in the received signal that would make it appear as though the local oscillator were changing. Most stations operate with a near perfect record, but they do make mistakes. This happens just often enough so that the user needs additional information to help him. This information takes the form of monthly or weekly notices of the actual phase of the signal. This is usually in terms of micro-

seconds advance or retard. A sample of NBS data for WWVB and other monitored stations is shown in figure 6.2. U. S. Naval Observatory (USNO) data for its stations are shown in figure 6.3. These data can be obtained free upon request.

The implication of possible transmitting errors to the user is that he cannot hurry through calibrations! Certain practices with respect to LF radio calibrations are highly recommended. These include careful attention to such details as receiver adjustments--do not change knob settings or cabling. Always operate the equipment continuously if possible. This gives you a continuous chart record so that you know when to expect sunrise/sunset phase shifts. You can also detect local interference and noise conditions. If your receiver has a mechanical phase counter, jot down the readings on the chart. This is a great help in trying to reconstruct events that happen on weekends or at night.

A typical occurrence for a tracking receiver is the cycle phase shift. Since the receiver is faithfully following a zero crossing of the received phase, it doesn't know which of the thousands of crossings it is locked to. If it loses lock, it will simply go to the next crossing that comes along. On a phase chart this will show up as a shift in phase equal to one cycle of the carrier.

For WWVB at 60 kHz, a cycle of phase equals 16.67 microseconds. If the chart recorder/receiver combination is producing a chart that is 50 microseconds across, the pen will move about 1/3 of the way across the chart for each cycle change. Such changes can be identified and must be ignored during calibration runs. This is due to the fact that phase charts are always ambiguous by an amount equal to the period of a carrier cycle.

How does the user tell whether the recorded phase difference indicates a change in his oscillator, the path, or the transmitted

3. PHASE DEVIATIONS FOR WWVB AND OTHER NBS-MONITORED BROADCASTS

WWVB (60 kHz)

Values given for WWVB are the time difference between the time markers of the UTC(NBS) time scale and the first positive-going zero voltage crossover measured at the transmitting antenna. The uncertainty of individual measurements is ± 0.5 microsecond. Values listed are for 1500 UTC.

Omega, North Dakota (13.1 kHz) and Omega, Hawaii (11.8 kHz)

Relative phase values are given for VLF stations and only the change from the previous available day's reading is published. Days when the data were satisfactory but readings were not taken (for example, on weekends or station maintenance days) are marked (-). If data are lost, continuity is also lost and the indication is (*), which means that reading cannot be compared to the previous day.

LORAN-C (DANA), Dana, Indiana (100 kHz) and LORAN-C (FALLON), Fallon, Nevada (100 kHz)

Values for Loran-C (Dana) and Loran-C (Fallon) are the time difference between the UTC(NBS) time pulses and the 1 pps output of the Loran-C receiver. Uncertainty in the measurements is ± 0.1 microsecond.

JULY	MJD	UTC(NBS) - WWVB (60 kHz) ANTENNA PHASE (in µs)	UTC(NBS) - RECEIVED PHASE (in µs)			
			OMEGA (11.8 kHz)	OMEGA (13.1 kHz)	LORAN-C (DANA) (100 kHz)	LORAN-C (FALLON) (100 kHz)
1	43690	5.90	(-)	(-)	(-)	(-)
	43691	5.99	(-)	(-)	(-)	(-)
3	43692	6.08	+ 2.5	+ 0.1	73684.84	3949.27
4	43693	5.98	(-)	(-)	(-)	(-)
5	43694	5.88	- 1.5	- 0.1	73685.21	3949.42
6	43695	5.83	+ 2.3	+ 0.1	73685.22	3949.47
7	43696	5.54	- 3.2	- 0.8	73685.27	3949.42
8	43697	5.61	- 8.8	(-)	(-)	(-)
9	43698	5.68	(-)	(-)	(-)	(-)
10	43699	5.75	- 9.6	+ 0.3	73685.41	3949.60
11	43700	5.93	+ 6.4	- 0.1	73685.40	3949.59
12	43701	5.66	+ 4.5	- 0.2	73685.41	3949.55
13	43702	5.81	+ 4.5	+ 0.5	73685.56	3949.67
14	43703	5.67	+ 0.6	0.0	73685.59	3949.69
15	43704	5.70	(-)	(-)	(-)	(-)
16	43705	5.72	(-)	(-)	(-)	(-)
17	43706	5.75	+ 0.2	+ 0.3	73685.69	3949.77
18	43707	5.66	+ 1.5	- 0.2	73685.75	3949.83
19	43708	5.74	- 3.1	+ 0.1	73685.80	3949.85
20	43709	5.79	- 0.5	- 0.1	73685.89	3949.81
21	43710	5.72	- 1.0	- 0.1	73685.92	3949.79
22	43711	5.78	(-)	(-)	(-)	(-)
23	43712	5.84	(-)	(-)	(-)	(-)
24	43713	5.90	+ 3.1	- 0.1	73685.93	3949.91
25	43714	5.86	- 0.8	- 0.1	73686.00	3949.96
26	43715	5.57	+ 1.0	+ 0.4	73685.96	3950.06
27	43716	5.71	+ 0.3	+ 0.2	73686.11	3950.15
		5.78		3	736	

FIGURE 6.2.   NBS DATA PUBLISHED IN NBS TIME AND FREQUENCY BULLETIN.

U. S. NAVAL OBSERVATORY
WASHINGTON, D.C. 20390

26 JUL. 1979

DAILY PHASE VALUES AND TIME DIFFERENCES SERIES 4

NO. 651

THE TABLE GIVES: UTC(USNO MC) - TRANSMITTING STATION

UNIT = ONE MICROSECOND

FREQUENCY KHZ (UTC)	MJD	LC/9970 NW PAC. 100	LC/4990 CEN. PAC. 100	LC/9930 E.COAST 100	LC/7970 NOR. SEA 100	LC/7990 MED. SEA 100	LC/7930 N. ATL. 100	LC/9990 N. PAC. 100	LC/9940 W.C.USA 100
JUL. 16	44070	-1.2	0.7	2.93	-2.2	0.2	-1.8	-1.5	-5.3
17	44071	-1.3	0.5	2.93	-2.4	0.1	-1.7	-1.5	-5.4
18	44072	-1.3	0.5	2.88	-2.4	0.1	-1.9	-1.3	-5.4
19	44073	-1.4	0.5	2.86	-2.4	0.1	-1.8	-1.4	-5.4
20	44074	-1.4	0.5	2.85	-2.5	0.0	-1.9	-1.5	-5.3
21	44075	-1.3	0.6	2.87	-2.3	0.2	-2.0	-1.8	-5.3
22	44076	-1.4	0.6	2.83	-	-	-2.1	-1.8	-5.3
23	44077	-1.3	0.6	2.90	-2.2	0.3	-2.2	-1.8	-5.3
24	44078	-1.4	0.6	2.88	-2.2	0.4	-2.0	-1.9	-5.3
25	44079	-1.3	-	2.88	-	-	-1.9	-	-5.3

FREQUENCY KHZ (UTC)	MJD	LC/5990 W.C.CAN 100	LC/7960 GULF AK 100	3 O/H 11.8 25MS+	1 O/ND 13.1 6MS+	7 O/ND 13.6 6MS+	5 O/A 13.6 30MS+	6 O/L 13.6 25MS+	4 O/T 13.6 11MS+	2 GBR 16.0 19MS+	8 NLK 18.6 12MS+
JUL. 19	44073	2.2	1.8	954	425	421	576	141	622	515	335
20	44074	2.2	1.8	954	425	421	576	142	622	514	341(1)
21	44075	2.1	1.5	9[illegible]	424	421	576	143	622	514	340
22	44076	-	1.6	953	425	421	574	143	622	514	340
23	44077	2.1	1.7	953	425	421	575	141	622	515	340
24	44078	-	1.7	953	426	421	575	141	622	514	340
25	44079		-	954	425	421	575	141	622	514	339

NOTES:

(1) NLK 19 JUL. AFTER MAINTENANCE PERIOD STEP ABOUT PLUS 6

(2) PROPAGATION DISTURBANCES WERE OBSERVED NEAR THE FOLLOWING TIMES:
20 JUL. 1940/5
21 JUL. 1110/4, 1340/6
22 JUL. 1310/5, 1440/5, 1735/4, 1925/5, 2110/6
23 JUL. 1435/4, 1645/5, 2055/5
24 JUL. 1[illegible]/5
25 JUL. 1600/4, 1735/3.

(3) OMEGA STATIONS OFF-AIR TIMES (FROM COAST GUARD ONSOD, WASHINGTON, DC):

JAPAN	17 JUL.	0039 TO 0041 UT, 0214 TO 0217 UT, 0232 TO 0234 UT, 0918 TO 0920 UT
LIBERIA	17 JUL.	0240 TO 0242 UT, 0305 TO 0403 UT
	21 JUL.	1620 TO 1622 UT
LA REUNION	17 JUL.	0423 TO 0425 UT, 0717 TO 0731 UT, 0735 TO 0737 UT, 1136 TO 1159 UT, 1247 TO 1306 UT, 1412 TO 1415 UT
	18 JUL.	0030 TO 0032 UT, 0632 TO 0645 UT, 1036 TO 1045 UT
	19 JUL.	0509 TO 0512 UT, 1059 TO 1105 UT, 1439 TO 1441 UT

FIGURE 6.3. USNO PHASE DATA, PUBLISHED WEEKLY.

signal? The answer is that he knows from experience! If you plan to use VLF or LF signals for frequency calibrations, you must become very familiar with the characteristics of the signals you are using. For accurate calibrations, you must obtain the station data from NBS or the USNO. Battery backup for receivers is highly desirable; in fact, almost necessary if you plan to calibrate precise oscillators over a period of many days. All of this may sound like a hopeless task, but it isn't. The results are worth the effort when you consider that you can perform calibrations up to 1 part in 10^{11} in 24 hours.

6.4 FIELD STRENGTHS OF VLF-LF STATIONS

The field strength of the WWVB signal has been measured along nine radial paths from the station. These measurements are summarized on the field contour map shown in figure 6.4. Not shown is the nonlinear field gradient between the 500 microvolt per meter contour and the 100 microvolt per meter contour.

Destructive interference occurs between the first hop skywave and the groundwave at

approximately 1200 km (750 miles) from the station. On some radial paths, this dip in field intensity is quite severe and has proven to be the cause of signal loss at certain times. The distance between the station and this null varies from day to night. It is also seasonal. The sharpness of the null is much less pronounced in the winter. The field intensity, in general, is slightly higher during the winter months. Shown in figures 6.5 and 6.6 are the field strengths along radials to Brownsville, Texas and Nantucket, Massachusetts, both measured in September. Other radial plots are available to interested users.

The signal strength of the U. S. Navy stations is usually very high unless they are undergoing repairs. The Omega navigation stations have a high power output also. But Omega is time shared; that is, the eight transmitters take turns broadcasting on the same frequencies. This has the effect of reducing the available signal power and commutation is required; that is, you must have a means of turning the receiver on only when the station you want to track is transmitting.

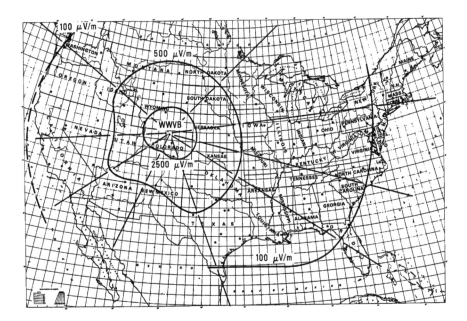

FIGURE 6.4. MEASURED FIELD INTENSITY COUNTOURS: WWVB @ 13 kW ERP.

111

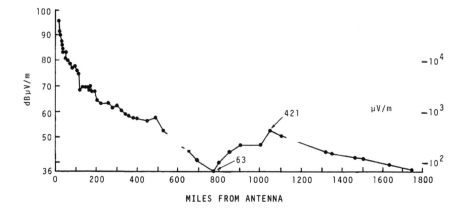

FIGURE 6.5. MEASURED FIELD INTENSITY, NANTUCKET, MASSACHUSETTS RADIAL, WWVB 60 kHz.

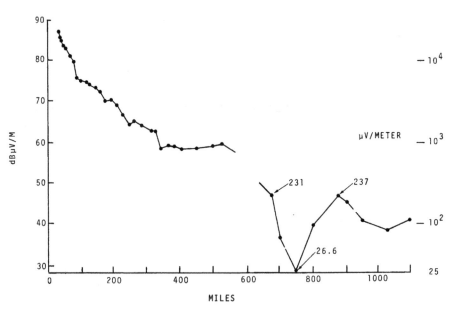

FIGURE 6.6. MEASURED FIELD INTENSITY, BROWNSVILLE, TEXAS RADIAL, WWVB 60 kHz.

6.5 INTERFERENCE

Some WWVB users along the U. S. Atlantic coast have experienced interference from a British standard frequency radio station, MSF. Some years ago, MSF was not only lower powered, but it was also offset in frequency from WWVB. However, it now radiates 25 kW of power on 60 kHz. There are various ways to tell if one is receiving MSF instead of WWVB. The 45° phase shift (discussed in Section 6.6.1, fig. 6.9) should be quite apparent if WWVB is being received and phase-tracked.

One solution is to use a directional antenna. If a loop is being used, the MSF signal can be nulled. Eastern U. S. receiving locations from Boston south that are using loops should be able to null MSF and still receive WWVB. Another method is to produce a unidirectional antenna by combining the voltages induced in a loop and a whip. If these two voltages are equal and in phase, the resultant pattern is a cardioid. Since these voltages are induced into the whip by the electric field, and into the loop by the magnetic field, the phase of the whip voltage must be shifted by 90° before combination. Interested readers are advised to consult a suitable antenna manual or to discuss their problems with the receiver manufacturer.

6.6 USING WWVB FOR FREQUENCY CALIBRATIONS

Our discussion of calibrations using low-frequency signals begins with WWVB (60 kHz). Frequency calibrations usually involve phase

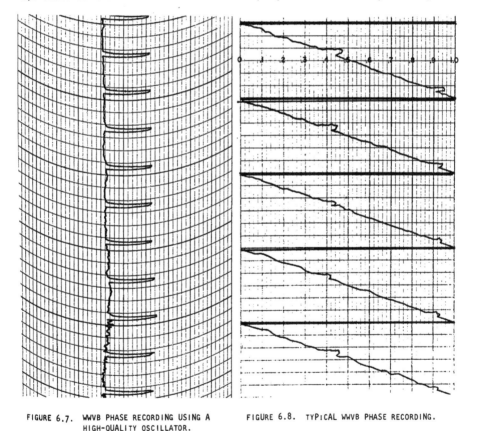

FIGURE 6.7.   WWVB PHASE RECORDING USING A   FIGURE 6.8.   TYPICAL WWVB PHASE RECORDING.
HIGH-QUALITY OSCILLATOR.

113

recordings of the received signals. Typically, a special tracking receiver is used to compare a local oscillator with the received WWVB signals. The output of the receiver can be displayed on its front panel digitally or as a meter reading. However, this discussion is limited to making a phase record on a chart recorder. Often, these recorders are built into the receivers. Since we are looking at a phase recording of the received signal, we will see the effect of any path changes. In addition, we will see our oscillator phase change relative to the WWVB signal.

The amount of phase change depends on the relative difference between our oscillator and the cesium oscillator controlling the WWVB signal at the transmitter. The sample phase chart shown in figure 6.7 has very little oscillator drift. This is because a quality oscillator was being measured. A more typical record would have a whole series of lines crossing the chart as in figure 6.8. This poses a dilemma for the user. If the oscillator being calibrated is far off frequency from WWVB, he needs either a fast chart or one that displays many microseconds of phase across the width of paper. For high-quality frequency sources, the paper can be slowed down or the chart width, in microseconds, can be reduced. The user is encouraged to study the manuals for the particular receiver he is using. If the user is planning to buy a receiver, he should consider the problem of what the record is going to look like before selecting a particular receiver/recorder combination.

Measurements made with VLF-LF signals should be done when the path from transmitter to receiver is all in sunlight. An all dark path would be second choice. The sunrise/sunset phase shifts (called diurnals) will affect the accuracy of calibrations and measurements should not be made during these times. There are other sudden phase shifts due to particles entering the ionosphere that can affect the signal. These usually last only ten minutes or a half-hour. This explains the advantage of having a chart that runs at least several hours for a calibration. During sudden phase shifts, the signal level may drop. Most receivers have a meter for signal strength, but the more elaborate receivers are able to chart record the signal amplitude. This is a very useful record if the calibrations being made are to be within parts in 10^{10} or better. In other words, you need all the help you can get if you are to make the very best measurements.

As mentioned earlier, some receivers use front panel displays to indicate frequency offset. Later in this chapter, we mention methods that use oscilloscopes for frequency calibration. Even a zero-beat or Lissajous

carrier to provide a frequency reference for calibration purposes. In order for the user to identify WWVB by its phase track alone, a 45° phase shift is programmed into the transmitted phase. This shift occurs at 10 minutes past the hour and returns to nominal value at 15 minutes past the hour. Three hours of strip chart at a typical user site might look something like figure 6.9.

6.6.2 METHODS OF FREQUENCY COMPARISON

Having listed some of the problems in LF propagation, we can now turn our attention to actual calibration methods. There are two ways to compute relative frequency from the plot in figure 6.1. One can work with the slope of the daytime plot--the straight portion--or one can take the single values at discrete points one day apart. On the figure, the 2200 UTC values for June 3 and June 4 are subtracted to give a value of 1 microsecond change for the 24-hour period. The relative frequency, $\Delta f/f$, is the same as the time offset, $\Delta t/T$. In this case:

$$\frac{\Delta f}{f} = \frac{\Delta t}{T} = \frac{1 \text{ µs}}{1 \text{ day}} = \frac{1 \text{ µs}}{86,400 \times 10^6 \text{ µs}}$$

$$= 1.16 \times 10^{-11}.$$

When the relative frequency is large, a useable value can be computed using only a few hours of the daytime slope. If the relative frequency is very small, many days may have to be used in the computation. This is especially true when the radio propagation stability is poor or the signal is noisy. If many days are involved in the measurement, cycle jumps and transmitter phase shifts entered into the phase plot must be carefully taken into account.

If the receiving site has no phase-tracking receiver available, other schemes may be used. If a tuned radio frequency receiver is available, the actual 60 kHz wave may be observed on an oscilloscope. If the oscilloscope is triggered by a local clock, the position on the oscilloscope face of a particular cycle will change with time. The amount of the change in any given time will provide a value of $\Delta t/T$ from which the relative frequency may be computed as before. A standard reference 60 kHz or submultiple, if available, may be used to trigger the oscilloscope.

Electronic counter techniques can also be used for frequency calibrations. Without going into detail, these techniques involve the use of the received signal as one input to a counter while the oscillator being calibrated is used as the other. Manufacturers of receivers and counters discuss these techniques in their manuals and application notes. You may also refer to chapters 1 and 4 of this book, which talk about counters and their uses. Keep in mind that the accuracies you can get with oscilloscopes and counters will generally be much less than those obtained by using tracking receivers. The cost will be less too. Also, a quick calibration on a counter cannot identify sudden phase changes, and you lose the big advantage of having a chart record of the station performance.

6.7 USING OTHER LF AND VLF STATIONS FOR TIME AND FREQUENCY CALIBRATIONS

6.7.1 THE OMEGA NAVIGATION SYSTEM

The Omega Navigation System is composed of a group of VLF radio stations operating in the 10 to 15 kHz range. Each station time-shares common frequencies used for navigation. In addition, each station may transmit some frequencies unique to that particular station. See Table 6.1 for details.

If one wishes to use an Omega station for frequency calibration, a phase-tracking receiver is highly recommended. If one of the navigation frequencies is to be used, then an Omega commutator must also be used. This is a device that turns the phase-tracking receiver on and off at the proper times to receive only the desired Omega station.

The frequencies and the format segments of the Omega stations are derived from cesium beam oscillators. The USNO monitors and reports the Omega stations' phase values. These stations radiate a nominal 10 kW of power. This power level should be sufficient to allow the user to receive at least three stations no matter where he is located.

A. Operating Characteristics of Omega

Omega transmitting stations operate in the internationally allocated VLF navigational band between 10 and 14 kHz. This very low transmitting frequency enables Omega to provide adequate navigation signals at much longer ranges than other ground-based navigation systems.

The operating characteristics of the system can be categorized as follows: signal format and control, the requirement for lane identification typical of phase comparison

TABLE 6.1. CHARACTERISTICS OF THE OMEGA NAVIGATION SYSTEM STATIONS

STATION	LOCATION	LATITUDE, LONGITUDE	POWER (kW)	ANTENNA	CARRIER FREQUENCIES*	ACCURACY
OMEGA Ω/N	ALDA, NORWAY	66° 25' N 13° 09' E	10	OMNI-DIRECTIONAL	10.2 A 11-1/3 C 13.6 B	5×10^{-12}
OMEGA Ω/L	MONROVIA, LIBERIA	06° 18' N 10° 40' W	10	OMNI-DIRECTIONAL	10.2 B 11-1/3 D 13.6 C	1×10^{-12}
OMEGA Ω/H	HAIKU, OAHU, HAWAII	21° 24' N 157° 50' W	10	OMNI-DIRECTIONAL	10.2 C 11-1/3 E 13.6 D	1×10^{-12}
OMEGA Ω/ND	LA MOURE, NORTH DAKOTA	46° 22' N 98° 20' W	10	OMNI-DIRECTIONAL	10.2 D 11-1/3 F 13.6 E	1×10^{-12}
OMEGA Ω/LR	LA REUNION	20° 58' S 55° 17' E	10	OMNI-DIRECTIONAL	10.2 E 11-1/3 G 13.6 F	1×10^{-12}
OMEGA Ω/A	GOLFO NUEVO, ARGENTINA	43° 03' S 65° 11' W	10	OMNI-DIRECTIONAL	10.2 F 11-1/3 H 13.6 G	1×10^{-12}
OMEGA Ω/T	TRINIDAD**	10° 42' N 31° 38' W	1	OMNI-DIRECTIONAL	10.2 B 11-1/3 D 13.6 C	1×10^{-12}
OMEGA Ω/J	TSUSHIMA IS., JAPAN	34° 37' N 129° 27' E	10	OMNI-DIRECTIONAL	10.2 H 11-1/3 B 13.6 A	1×10^{-12}

*SEE FIGURE 6.10 FOR TRANSMISSION SCHEDULE OF VARIOUS FREQUENCIES.
**TRINIDAD IS A TEMPORARY STATION. PERMANENT STATION WILL BE IN AUSTRALIA.

systems and the process of handling errors attributable to signal propagation.

All stations now transmit three basic navigational frequencies (10.2 kHz, 11-1/3 kHz, 13.6 kHz) more or less omnidirectionally. In order to prevent interference, transmissions from each station are time-sequenced as shown in figure 6.10.

This pattern is arranged so that during each transmission interval (approximately 1 second), only three stations are radiating, each at a different frequency. The duration of each transmission varies from 0.9 to 1.2 seconds, depending on the station's assigned location within the signal pattern. With eight stations in the implemented system and a silent interval of 0.2 second between each transmission, the entire cycle of the signal pattern repeats every 10 seconds.

Besides the three basic navigational frequencies, other frequencies have been added to the Omega signal format. Original plans were made to transmit two unique frequencies at each station for the purpose of inter-station time synchronization but this requirement has been removed through use of highly stable cesium frequency standards.

Present plans call for the incorporation of a fourth navigation frequency, 11.050 kHz, which will allow for lane resolution capability as great as 288 nautical miles. In addition, a unique frequency transmission for each station can be added which will aid in time dissemination by providing a beat frequency and a high duty cycle at that frequency. The changes depicted in figure 6.10, which are not yet implemented, should be made by 1979.

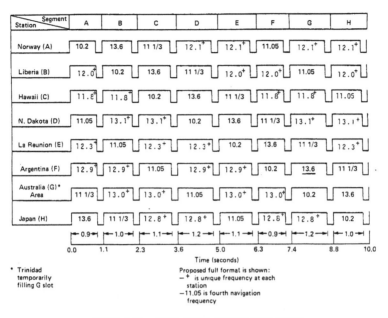

Segment Station	A	B	C	D	E	F	G	H
Norway (A)	10.2	13.6	11 1/3	12.1⁺	12.1⁺	11.05	12.1⁺	12.1⁺
Liberia (B)	12.0⁺	10.2	13.6	11 1/3	12.0⁺	12.0⁺	11.05	12.0⁺
Hawaii (C)	11.8⁺	11.8⁺	10.2	13.6	11 1/3	11.8⁺	11.8⁺	11.05
N. Dakota (D)	11.05	13.1⁺	13.1⁺	10.2	13.6	11 1/3	13.1⁺	13.1⁺
La Reunion (E)	12.3⁺	11.05	12.3⁺	12.3⁺	10.2	13.6	11 1/3	12.3⁺
Argentina (F)	12.9⁺	12.9⁺	11.05	12.9⁺	12.9⁺	10.2	13.6	11 1/3
Australia (G)* Area	11 1/3	13.0⁺	13.0⁺	11.05	13.0⁺	13.0⁺	10.2	13.6
Japan (H)	13.6	11 1/3	12.8⁺	12.8⁺	11.05	12.8⁺	12.8⁺	10.2

|←0.9→| |←1.0→| |←1.1→| |←1.2→| |←1.1→| |←0.9→| |←1.2→| |←1.0→|

0.0   1.1   2.3   3.6   5.0   6.3   7.4   8.8   10.0

Time (seconds)

* Trinidad
  temporarily
  filling G slot

Proposed full format is shown:
— ⁺ is unique frequency at each
   station
— 11.05 is fourth navigation
   frequency

FIGURE 6.10. OMEGA SIGNAL TRANSMISSION FORMAT.

B. Synchronization Control

The Omega signal format is designed so that each station within the network can be identified by the transmission of a particular frequency at a prescribed time. In addition, the synchronization of all transmissions is tightly controlled and the phase relationships between all signals are maintained to within a few centicycles. With this high phase stability in the transmissions, the accuracy of a navigational fix is then primarily limited to the receiver and the accuracy of the navigator's propagation correction tables.

All Omega transmitting stations are synchronized by means of very stable cesium beam frequency standards. These standards or clocks are referenced to the atomic time scale which differs from Coordinated Universal Time (UTC) more commonly in use. Thus, in 1978, the Omega epoch or time reference is seven seconds ahead of UTC since the yearly adjustments for earth motion have not been made to make Omega Epoch in agreement with UTC.

C. Propagation Characteristics

The propagation characteristics that permit the use of Omega at great range also introduce certain limitations. Two areas that require special attention are normal time variations and modal interference. Since signals are propagated within the waveguide formed by the earth and ionosphere, changes in propagation parameters such as velocity may be expected as a result of changes in the ionosphere or ground.

The most obvious navigationally undesirable variation is a daily or diurnal phase change. Normal changes in illumination of the ionosphere by the sun throughout the day may cause an uncorrected phase measurement to vary as much as one complete cycle. Since these variations are highly repeatable, prediction and correction are possible.

Unpredictable short-term variations may also occur. Ninety-five percent of the time, these are small variations related to random propagational variations whcih will not degrade normal navigational accuracy. Occasionally, however, large disturbances can occur as a result of solar emission of X-ray or particle bursts. The emission of X-rays from the sun occasionally cause a short-term disruption of Omega signals which is referred to as a Sudden Phase Anomaly (SPA). The duration of a SPA is generally not greater than one hour but an LOP may experience a shift of several miles. SPA's occur with an average frequency

117

of about 7-10 per month. They usually affect signals from only a few stations at a time since X-rays from the sun tend to enter a limited illuminated portion of the earth's surface.

The release of a large quantity of protons from the sun, although an infrequent occurrence, produces what is known as a Polar Cap Disturbance (PCD). The effect of a PCD may be to shift an LOP 6-8 miles for a period of several days. This disturbance generally lasts for several days and varies in magnitude during the period. PCD's affect only those transmissions involving arctic propagation paths. Because of its possible long duration and large LOP shift, PCD notices are broadcast as navigational warning messages.

Modal interference is a special form of signal interference wherein the various waveguide modes of signal propagation interfere with each other and irregularities appear in the phase pattern. Ideally, one mode would be completely dominant at all times and the resultant phase grid would be regular. In practice, competing modes do not completely disappear and three situations are recognizable:

1. If the competing mode is very small, then the dominant mode will establish a nearly regular phase pattern as is intended, and usually this is what happens during the day.

2. A second possibility is that the competing mode may be almost equal to the dominant mode.

3. The potentially serious case is that in which modal dominance can change. This may occur, for example, if one mode is dominant during the day and a second mode at night. Clearly, somewhere during sunset and sunrise, the transitional period, the two modes must be equal. Depending upon phasing of the modes at equality, abnormal transitions may occur in which cycles are "slipped" or lost. Positional errors of a full wavelength are possible under such conditions and use of a station so affected should be avoided. If this is not possible, particular attention must be given to proper lane identification.

Propagation Corrections (PPC's) must be applied to each Omega receiver reading to compensate for ionospherically induced signal variations and thereby improve position fixing accuracy. Omega propagation correction tables for each transmitting station (A through H) contain necessary data for correcting Omega receiver readouts affected by prevailing propagation conditions relative to the nominal conditions on which all charts and tables are based. A brief introduction, which also describes the arrangement and application of the corrections together will illustrative examples, precedes the tabular data within each PPC table. For more detailed information on the Omega Charts and Tables as well as information on obtaining these charts, write to: Defense Mapping Agency, Hydrographic Center, Attn: Code DSI-2, Washington, DC. 20390.

D. Omega Notices and Navigational Warnings

As with other navigational aids, information is disseminated concerning station off-the-air periods, VLF propagation disturbances, and other pertinent data affecting usage of the system. Status information on Omega stations can be obtained on the telephone by calling (202) 245-0298 in Washington, D. C.

Omega status messages are also broadcast by various time service stations. NBS stations WWV and WWVH transmit a 40-second message at 16 minutes and 47 minutes after the hour, respectively. Norway Radio Station Rogaland broadcasts notices in international morse code on HF 4 times daily. Other stations in the global time service and maritime information network are expected to be added in the future.

In addition to the regular issued notices on station off-air periods, major planned maintenance may take place in the months listed for each station:

ARGENTINA	–	MARCH
LIBERIA	–	APRIL
HAWAII	–	MAY
LA REUNION	–	JUNE
NORWAY	–	JULY
TRINIDAD	–	FEBRUARY
NORTH DAKOTA	–	SEPTEMBER
JAPAN	–	OCTOBER

The actual off-air times are disseminated as noted above sufficiently in advance and may vary from a few days to several weeks depending on the maintenance or repairs required.

6.7.2 DCF 77, WEST GERMANY (77.5 kHz)

DCF 77 is located at Mainflingen, about 25 kilometers southeast of Frankfurt/Main. It is operated by the Physikalisch-Technische Bundesanstalt (PTB) in Braunschweig.

118

DCF 77 transmits Central European Time
(ET) or Central European Summer Time (CEST),
¡ich is equal to UTC(PTB) plus 1 or 2 hours
spectively. The carrier phase is controlled
th respect to UTC(PTB); phase time changes
'e kept smaller than 0.5 µs. The station
·ansmits continuously except for short inter-
¡ptions because of technical faults or main-
¡nance. Longer breaks may be experienced
¡ring thunderstorms.

A. Time Signals

The carrier is modulated by means of
¡conds markers with the exception of second
¡mber 59 of each minute which signifies that
¡e next marker will be the minute marker. At
¡e beginning of each second (with the excep-
on of the 59th) the carrier amplitude is
¡duced to about 25% for a duration of 100 or
200 ms. The beginning of the decrease of the
carrier amplitude characterizes the exact
beginning of the corresponding second.

The seconds markers are phase-synchronous
with the carrier. In general, the uncertainty
of the received DCF 77 time signals is large
compared to that of the emitted time signals.
This results from the limited bandwidth of the
transmitter antenna, skywave influence and
possible interference. At a distance of some
hundred kilometers, a time signal uncertainty
of less than 0.1 ms is achievable.

B. Time Code

The seconds marker durations of 100 and
200 ms correspond to binary 0 or 1, respectively, in a BCD code used for the coded
transmission of time and date.

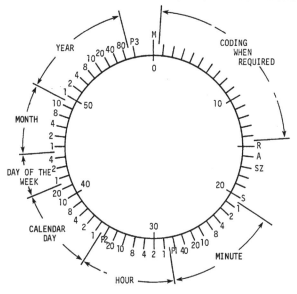

- (second marker 20), a 200 ms marker,
  designates the start of the time
  information.

Z - Summer Time bit.

- Announcement of an approaching change
  from CET.

R - Second marker 15 has a duration of 200 ms
    if the reserve antenna is used. (This may
    result in a small phase time change of the
    carrier due to the different location of
    the antenna.

P1, P2, P3 - Parity check bits.

FIGURE 6.11. DCF 77 TIME CODE FORMAT.

There are three groups of time information, each followed by a parity check bit P:

P1 = NUMBER OF THE MINUTE

P2 = NUMBER OF THE HOUR

P3 = NUMBERS OF THE CALENDAR DAY,
THE DAY OF THE WEEK, THE
MONTH AND THE YEAR

Counting the binary "ones" of the information group concerned and of the corresponding parity check bit yields an even number.

In the case of the transmission of Legal Time in the form of CEST, seconds marker number 17 has a duration of 200 ms. During one hour before the change from CET to CEST or from CEST to CET, seconds marker 16 has a duration of 200 ms, thus announcing the approaching change. The coding is shown in figure 6.11.

General information about DCF 77 is shown in Table 6.2.

6.7.3 HBG, SWITZERLAND (75 kHz)

HBG in Prangins, Switzerland is operated by the Observatoire Cantonal in Neuchatel.

The useful range for clock synchronization is 3000 kilometers.

A. Signal Format

The carrier is interrupted for 0.1 second at the beginning of each second. An additional interruption occurs from 0.2 to 0.3 second at the beginning of each minute.

B. Time Code

No continuous time code is transmitted. During the weekends of changes to daylight saving time or back to standard time, a special code is transmitted, including a separate 4-bit address for each European country. Additional details about HBG are shown in Table 6.2.

6.7.4 JG2AS/JJF-2, JAPAN (40 kHz)

The Radio Research Laboratories in Tokyo operate JG2AS and JJF-2. JJF-2 consists of telegraph signals; JG2AS broadcasts in the absence of these telegraph signals. The UTC time scale is broadcast. The call sign is transmitted twice each hour (at 15 and 45 minutes) in Morse code.

(1) Hourly modulation schedule

Call sign: Three times by the Morse code

(2) Wave form of second pulses

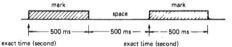

(3) Identification of minute signal

FIGURE 6.12.   JG2AS EMISSION SCHEDULE.

120

The 40 kHz carrier frequency (of 500 ms duration) is broadcast at the beginning of each second except for the 59th pulse in each minute, which is 100 ms long.

The JG2AS emission schedule is shown in figure 6.12. Other details are shown in Table 6.2.

6.7.5 MSF, ENGLAND (60 kHz)

MSF is located at Rugby, United Kingdom. It is operated by the National Physical Laboratory in Teddington, Middlesex. Transmissions are continuous except for a maintenance period from 1000-1400 UTC on the first Tuesday of each month. MSF transmits both a fast code and a slow code.

A. Fast Code

Calendar information; i.e., month, day, hour, and minute are given, the complete data burst occupying a total of 330 ms at the fast bit rate of 100 bits/second in the 500 ms interval following the minute marker. The full format of the code is shown in figure 6.13. It includes a control pulse and marker bit to initiate recognition systems in the receiver, 24 data bits, 2 parity checks (odd parity) and a British Standard Time (BST) indicator (UTC + 1 hour) so transmissions are advanced when BST is in effect.

B. Slow Code

The slow code operates at a 1 bit/second rate. The sequence of signals in each minute of the transmission is shown in figure 6.14. Pulse length modulation is used, the carrier being interrupted for 100 ms for a logic "0" and for 200 ms for a logic "1." Full calendar and time information is provided; i.e., year, month, day-of-month, day-of-week, hour, and minute referred to the minute marker immediately following the code. The minute itself is identified by an 8-bit precursor transmitted in seconds 52 to 59 in the form of the sequence 0111 1110. Exhaustive trials by computer simulation have confirmed that this sequence is unique and will not occur as part of the code proper at any time; the identifier is also well protected against a code with one or two errors present.

The introduction of a minute identifier enables a very simple form of receiver/decoder to be developed making use of a shift register whose length can be adjusted to acquire as little or as much of the code as is desired. For example, if the register is only 8 bits long, this will be sufficient to identify each minute but will exclude all the other data, which will pass through the shift register but will not be recognized since the register responds only to the combination 0111 1110. When this is received, the next carrier interruption is identified as the minute marker. Similarly a register 21 bits long will extract hour and minute information while the maximum length of 43 bits is necessary for a full decoding of calendar and time-of-day information.

The code described so far contains the basic, primary information at what may be called the first level of coding. However, further subsidiary information is also encoded at a second level where logic "1" is indicated by a carrier interruption in the interval 200 to 300 ms following a seconds marker. This form of coding, as shown in figure 6.14, is applied within the identifier at seconds 54 - 58, inclusive, and includes four parity checks and also a BST indicator. The latter indicates the operation of BST, as in the fast code, but with the difference that the clock generating the slow code will be set 1 hour ahead of UTC when British Summer Time is in effect. The second level of coding also embraces the DUT1 indication in seconds 1-16, inclusive, and brings this information into the same format as the other signals.

The adoption of a minute identifier immediately preceding the minute creates difficulties in introducing a leap second since a positive leap second would be identified as the following minute marker while a negative leap second would remove the last bit in the identifier itself. When necessary, therefore, the leap second is inserted following second 16 where it can be accommodated without disturbance to the coded sequence and identifier.

Additional details about MSF can be found in Table 6.2.

6.7.6 U.S. NAVY COMMUNICATION STATIONS

Some of the Navy communication stations are commonly used to steer frequency sources: NAA, NDT, NLK, NPM, NSS, and NWC. However, use of these stations for frequency calibration is strictly secondary to their primary function of military communications. These stations are subject to frequency changes, outages, and changes in schedule. Changes are announced in advance to interested users by the USNO. All stations operate

121

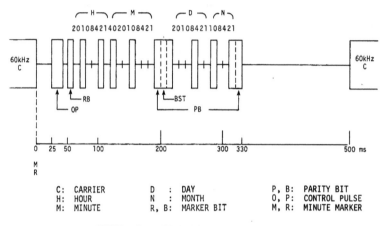

C:	CARRIER	D :	DAY	P, B:	PARITY BIT
H:	HOUR	N :	MONTH	O, P:	CONTROL PULSE
M:	MINUTE	R, B:	MARKER BIT	M, R:	MINUTE MARKER

FIGURE 6.13. MSF 60 kHz FAST CODE FORMAT.

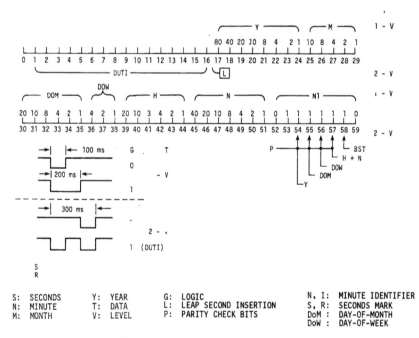

S:	SECONDS	Y:	YEAR	G:	LOGIC	N, I:	MINUTE IDENTIFIER
N:	MINUTE	T:	DATA	L:	LEAP SECOND INSERTION	S, R:	SECONDS MARK
M:	MONTH	V:	LEVEL	P:	PARITY CHECK BITS	DoM :	DAY-OF-MONTH
						DoW :	DAY-OF-WEEK

FIGURE 6.14. MSF 60 kHz SLOW CODE FORMAT.

122

continuously except for maintenance periods, which are shown below:

STATION	MAINTENANCE SCHEDULE
NAA	1400-1800 UTC each Monday; rescheduled for preceding Friday if Monday is a holiday. Half-power operations on Wednesday and Thursday from 1200-2000 UTC. Scheduled/unscheduled half-power operations as required.
NDT	2200-0600 UTC Thursday/Friday except first Thursday/Friday of each month, then 2200-0800 UTC.
NLK	1700-2200 UTC first & third Thursday of each month.
NPM	2200-0200 UTC each Thursday and Friday.
NSS	1200-2000 UTC every Tuesday.
NWC	0001-0400 UTC every Wednesday.

Other details of these stations are shown in Table 6.2.

6.7.7 OMA, CZECHOSLOVAKIA (50 kHz)

OMA in Liblice, Czechoslovakia, transmits the UTC time scale. The carrier is phase-locked over a distance of 30 kilometers to the phase of UTC (TP), the Czechoslovakian time scale. As a result, no long-term drift with respect to UTC(TP) can occur. The rms phase deviation is of approximately 100 ns; however, during strong winds the phase excursions may reach 500 ns since the antenna is not servo-controlled.

A.   Time Code

The format of the OMA time code is shown in figure 6.15. The signal is both amplitude and phase keyed. A1 seconds ticks of 0.1 second duration are prolonged to 0.5 second to identify minutes. A time code representing the time of day of CET (Central European Time = UTC + 1 hour) is transmitted by means of phase-reversing the carrier for 0.1 second in corresponding seconds of the current minute. Thus, in each minute a series of 4 pulses (P1 to P4) is transmitted.

The pulses, together with the A1 minute pulse (M) form 4 time intervals which, measured in seconds, represent 4 numbers: units of minutes (UM), tens of minutes (TM), units of hours (UH), and tens of hours (TH) of CET. As shown in figure 6.15, the duration of each interval is prolonged by 1 second so that the number 0 may be transmitted as a 1 second interval. In the illustrated case, 10 hours 01 minute of CET is transmitted which corresponds to the following intervals:

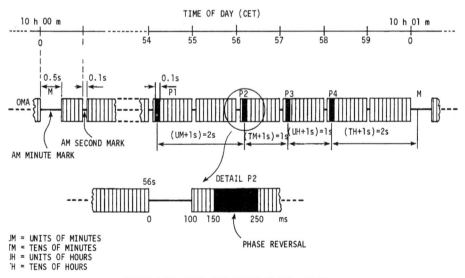

FIGURE 6.15.   TIME CODE FORMAT OF OMA - 50 kHz.

P2 - P1 (in seconds)   = UM + 1

P3 - P2   = TM + 1

P4 - P3   = UH + 1

M - P4   = TH + 1

Completion of the code by the date is planned for the near future.

Further details about OMA are shown in Table 6.2.

6.7.8 RBU (66-2/3 kHz) AND RTZ (50 kHz), USSR

Table 6.2 gives details about these stations. However, the contents of the UT1 corrections are described below:

The information about the value and the sign of the DUT1 + dUT1 difference is transmitted after each minute signal by the marking of the corresponding second signals by additional impulses. In addition, it is transmitted in accordance with CCIR code. Additional information on dUT1 is given, specifying more precisely the difference UT1 - UTC down to multiples of 0.02 second, the total value of the correction being DUT1 + dUT1. Positive values of dUT1 are transmitted by the marking of "p" seconds markers within the range between the 20th and 25th seconds so that dUT1 = +0.02 second x "p." Negative values of DUT1 are transmitted by the marking of "q" seconds markers within the range between the 35th and 40th seconds so that dUT1 = -0.02 second x "q."

6.7.9 VGC3 AND VTR3, USSR

These stations operate at 25.0, 25.1, 25.5, 23.0, and 20.5 kHz. Each station transmits 40 minutes 4 times/day:

VGC3 from 0036 to 0117, 0336 to 0417, 0636 to 0717, and 1736 to 1817 hours UTC;

VTR3 from 0536 to 0617, 1436 to 1517, and 1836 to 1917 hours UTC.

Two types of modulation signal are transmitted during a duty period: (1) A1 signals with carrier frequency 25 kHz, duration 0.0125, 0.025, 0.1, 1, and 10 seconds with repetition periods of 0.025, 0.1, 1, 10, and 60 seconds respectively; and (2) A0 signals with carrier frequencies 25.0, 25.1, 25.5, 23.0, and 20.5 kHz. The phases of these signals are matched with the time markers of the transmitted scale.

No UT1 information is transmitted. See Table 6.2 for further details about these stations.

124

TABLE 6.2. STANDARD FREQUENCY AND TIME SIGNAL BROADCASTS IN THE LOW FREQUENCY BAND

STATION	LOCATION LATITUDE LONGITUDE	POWER (kW)	ANTENNA	STANDARD FREQUENCIES USED CARRIER (kHz)	STANDARD FREQUENCIES USED MODULATION (Hz)	TIMES OF UT TRANSMISSIONS	ACCURACY	DUT1 CODE	NOTES
DCF77	MAINFLINGEN, GERMANY, F.R. 50° 01' N 9° 00' E	20	OMNI-DIRECTIONAL	77.5	1	CONTINUOUS*	5×10^{-13}	NONE	SEE SECTION 6.7.2 FOR SPECIFIC DETAILS.
GBR	RUGBY, UNITED KINGDOM 52° 22' N 1° 11' W	60	OMNI-DIRECTIONAL	16	1*	FROM 0255 to 0300, 0855 TO 0900, 1455 TO 1500, AND 2055 TO 2100 HOURS UT.	2×10^{-12}	CCIR CODE: DOUBLE PULSE	*A1 TELEGRAPHY SIGNALS.
HBG	PRANGINS, SWITZERLAND 46° 24' N 6° 15' E	10	OMNI-DIRECTIONAL	75	1	CONTINUOUS	2×10^{-12}	NONE	SEE SECTION 6.7.3 FOR SPECIFIC DETAILS.
JG2AS JJF-2	SANWA, SASHIMA IBARAKI, JAPAN 36° 11' N 139° 51' E	10	OMNI-DIRECTIONAL	40	1	CONTINUOUS	1×10^{-11}	NONE	SEE SECTION 6.7.4 FOR SPECIFIC DETAILS.
MSF	RUGBY, UNITED KINGDOM 52° 22' N 1° 11' W	25	OMNI-DIRECTIONAL	60	1	CONTINUOUS	2×10^{-12}	CCIR CODE: DOUBLE PULSE	SEE SECTION 6.7.5 FOR SPECIFIC DETAILS.
NAA	CUTLER, MAINE 44° 38' N 67° 17' W	1000 (FULL) 500 (HALF)	OMNI-DIRECTIONAL	17.8	NIL	NIL	1×10^{-11}	NONE	TRANSMISSION MODE: MSK. SEE SECTION 6.7.6 FOR SPECIFIC DETAILS.
NDT	YOSAMI, JAPAN 34° 58' N 137° 01' E	40	OMNI-DIRECTIONAL	17.4	NIL	NIL	1×10^{-11}	NONE	TRANSMISSION MODE: MSK. SEE SECTION 6.7.6 FOR SPECIFIC DETAILS.
NLK	JIM CREEK, WASHINGTON 48° 12' N 121° 55' W	1200 130	OMNI-DIRECTIONAL	18.6	NIL	NIL	1×10^{-11}	NONE	TRANSMISSION MODE: MSK. SEE SECTION 6.7.6 FOR SPECIFIC DETAILS.

NORTH WEST CAPE, AUSTRALIA 21° 49' S 114° 10' E	OMNI-DIRECTIONAL	NIL	$\times 10^{-11}$	TRANSMISSION MODE: MSK. SEE SECTION 6.7.6 FOR SPECIFIC DETAILS.
KHABAROVSK, USSR 40° 30' N 134° 51' E	OMNI-DIRECTIONAL	40 MINS 4 TIMES/DAY		SEE SECTION 6.7.9 FOR SPECIFIC DETAILS.
GERKY, USSR 56° 11' N 43° 58' E	OMNI-DIRECTIONAL	40 MINS 4 TIMES/DAY		SEE SECTION 6.7.9 FOR SPECIFIC DETAILS.

power 200 milliseconds later for a binary zero, 500 milliseconds later for a binary one, and 800 milliseconds later for a reference marker or position identifier. Certain groups of pulses are encoded to represent decimal numbers which identify the minute, hour, and day of year. The binary-to-decimal weighting scheme is 8-4-2-1 with the most significant binary digit transmitted first. Note that this weighting sequence is the reverse of the WWV/WWVH code. The BCD groups and their basic decimal equivalents are tabulated below:

BINARY GROUP	DECIMAL
WEIGHT: 8 4 2 1	EQUIVALENT
0 0 0 0	0
0 0 0 1	1
0 0 1 0	2
0 0 1 1	3
0 1 0 0	4
0 1 0 1	5
0 1 1 0	6
0 1 1 1	7
1 0 0 0	8
1 0 0 1	9

The decimal equivalent of each group is derived by multiplying the individual binary digits by the weight factor of their respective columns and then adding the four products together. For example, the binary sequence 1001 in 8-4-2-1 code is equivalent to $(1 \times 8) + (0 \times 4) + (0 \times 2) + (1 \times 1) = 8 + 0 + 0 + 1 = 9$, as shown in the table. If fewer than nine decimal digits are required, one or more of the high-order binary digits may be dispensed with.

Once every minute, in serial fashion, the code format presents BCD numbers corresponding to the current minute, hour, and day on the UTC scale. Two BCD groups identify the minute (00 through 59); two groups identify the hour (00 through 23); and three groups identify the day of year (001 through 366). When representing units, tens, or hundreds, the basic 8-4-2-1 weights are multiplied by 1, 10, or 100 respectively. The coded information refers to the time at the beginning of the one-minute frame. Within each frame, the seconds may be determined by counting pulses.

Every new minute commences with a frame reference pulse which lasts for 0.8 second. Also, every ten-second interval within the minute is marked by a position identifier pulse of 0.8-second duration.

UT1 corrections to the nearest 0.1 second are transmitted at seconds 36 through 44 of each frame. Coded pulses at 36, 37, and 38

127

seconds indicate the positive or negative relationship of UT1 with respect to UTC. Pulses at 36 and 38 seconds are transmitted as binary ones only if UT1 is early with respect to UTC, in which case a correction must be added to the UTC signals to obtain UT1. The pulse transmitted at 37 seconds is a binary one if UT1 is late with respect to UTC, in which case the required UT1 correction must then be subtracted. The magnitude of the UT1 correction is transmitted as a BCD group at 40, 41, 42, and 43 seconds. Because UT1 corrections are expressed in tenths of seconds, the basic 8-4-2-1 weight of that particular binary group is multiplied by 0.1 to obtain its proper decimal equivalent.

Figure 6.16 shows a sample frame of the time code in rectified or dc form. The negative-going edge of each pulse coincides with the beginning of a pulse. Position identifiers are labeled P_1, P_2, P_3, P_4, P_5, and P_0. Brackets show the demarcation of the minutes, hours, days, and UT1 sets. The applicable weight factor is printed beneath the coded pulses in each BCD group. Except for the position identifiers and the frame reference marker, all uncoded pulses are binary zeros.

In figure 6.16, the most significant digit of the minutes set is $(1 \times 40) + (0 \times 20) + (0 \times 10) = 40$; the least significant digit of that set is $(0 \times 8) + (0 \times 4) + (1 \times 2) + (0 \times 1) = 2$. Thus, at the beginning of the frame, UTC was precisely 42 minutes past the

hour. The sets for hours and days reveal further that it is the 18th hour of the 258th day of the year. The UT1 correction is -0.7 second, so at the beginning of the frame the correct time on the UT1 scale was 258 days, 18 hours, 41 minutes, 59.3 seconds.

6.9.2 TIME TRANSFER USING THE TRANSMITTED ENVELOPE

There are three ways to use the code. The first way--also the most expensive--is to use a WWVB receiver with an automatic decoder and display unit. There are a number of manufacturers that produce this type of equipment. A second level of user equipment consists of a receiver having internal logic circuits that provide a level shift code output. This code may be applied directly to a strip chart or to an oscilloscope for manual use. The simplest equipment might consist of a tuned radio frequency receiver with an oscilloscope used to observe the amplified signal.

Figure 6.17 shows the signal envelope as seen on an oscilloscope at the transmitter. The on-time points are easily seen. With the horizontal scale of the oscilloscope expanded to 200 microseconds/centimeter, the on-time point can be determined to about ± 2 cycles of the 60 kHz carrier. Of course, the farther the receiving site is from the transmitter, the more the signal-to-noise ratio degrades

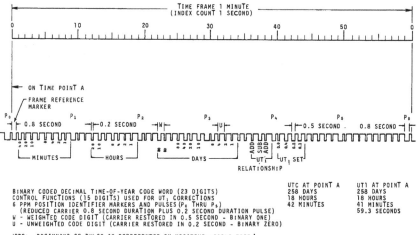

FIGURE 6.16. WWVB TIME CODE FORMAT

MAIN TRANSMITTER

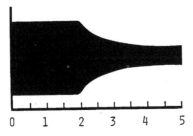

STANDBY TRANSMITTER

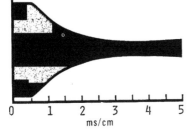

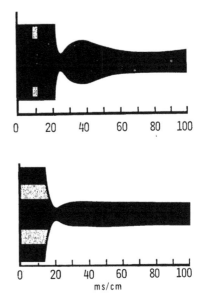

FIGURE 6.17. WWVB ENVELOPES AS TRANSMITTED.

until some averaging techniques must be employed. Since the WWVB data rate is so low--one bit per second--visual integration is very difficult, especially when the horizontal scale is expanded to allow greater time resolution. Averaging or signal integration using a signal averager or an oscilloscope camera works quite well at remote receiving locations. (See Chapters 4 and 5 for more information about these techniques.)

The actual process of time recovery consists of viewing the signal envelope on an oscilloscope that is being triggered by a local clock at the receiving site. Having previously determined the path delay, its value can be subtracted from the total observed delay on the oscilloscope to obtain the local clock error.

In very noisy locations, where the on-time point is difficult to identify, a photograph of the waveform can be helpful. Sometimes it is advantageous to allow a number of oscilloscope traces to be exposed on the same piece of film. This has an averaging effect that simplifies the location of the on-time point. However, a single exposure can

be used. Take the photo and draw one horizontal line through the average of the waveform amplitude and another line along the average slope of the dropout. The intersection of these lines is the on-time point.

Receiving sites that are long distances from the transmitter have an added problem when high resolution timing is desired. If the delay to the receiver site is, say, 15 milliseconds and the sweep speed is set to 1 millisecond/centimeter, the on-time point of the envelope will be off the oscilloscope face. This problem can be resolved if a second clock is available. The auxiliary clock can be adjusted late so that when used for the oscilloscope trigger source, the on-time point will be near the beginning of the trace. The sweep speed can now be increased to obtain the desired resolution. The time intervals to be accounted for are the time between the beginning of the trace and the on-time point, and the interval between the local clock and the auxiliary clock. The local clock error, Δt_e, will be

$$\Delta t_e = D - \Delta t_c - \Delta t_0 ,$$

129

where: D is the propagation delay

Δt_c is the interval between the two clocks

Δt_0 is the trace delay on the oscilloscope

If Δt_e is positive, then the local clock pulse occurs after the transmitted time pulse. If Δt_e is negative, the local clock pulse occurs first.

6.10 SUMMARY

Although you can get better accuracies by using LF and VLF broadcasts than you can from HF broadcasts, use of these signals poses special problems that are not encountered at HF. But if you need frequency accuracies of a part in 10^{11} or better, or time synchronization to 500 microseconds, VLF or LF broadcasts may suit your needs very well. However, you must be willing the spend the time, money, and effort necessary to get maximum results.

You do have an alternative, though. The frequency calibration service using network television and the TV Line-10 method of obtaining time and/or frequency may be less expensive and easier to use.

130

One of the newer frequency calibration services offered by NBS uses TV signals which have some advantages over radio broadcasts: First, the signals are readily available and usually very strong. Second, television receivers and antennas are simple to operate and install. Radio signals are carefully controlled at the transmitter, but when received by the user, they are sometimes degraded in quality and difficult to locate in the radio bands. Calibration via television lets the user concentrate on the calibration and spend less time fussing to get a good signal.

This frequency calibration service uses television signals in a transfer mode. If a user wants to make a calibration, he compares a TV signal coming from the national networks with his local oscillator. NBS monitors the same signal, calibrates it, and tells the user what correction to use.

The network signals are controlled by atomic oscillators and change their frequency very slowly, so the user needs new correction data at infrequent intervals. NBS checks the networks daily but publishes the corrections only once a month. This is often enough. The results obtained by this method of frequency calibration match or exceed any other method available. Anyone desiring a frequency calibration can make one quickly and have confidence in the results. Accuracies obtainable

range from a part in 10^9 to a few par 10^{11}.

7.1 HOW THIS FITS INTO OTHER NBS SER\

NBS operates radio stations WWV and WWVB. It also has a TV Line-10 which is explained in the next chapte service complements the radio broadca Line-10. Users who have the equipn skill necessary to use radio signa probably want to continue to use th ever, if TV signals are available radio services fall short in some then a switch is indicated. Many get a good TV signal in their area, it difficult to erect a suitable a the NBS radio signals. This telev quency calibration service operate well on cable (CATV) systems, the nels, and in areas of Canada and N receive U.S. network signals. B ceiver must be a color set.

7.2 BASIC PRINCIPLES OF T
FREQUENCY CALIBRATION

The major television network use atomic oscillators to genera erence signals. These are eith

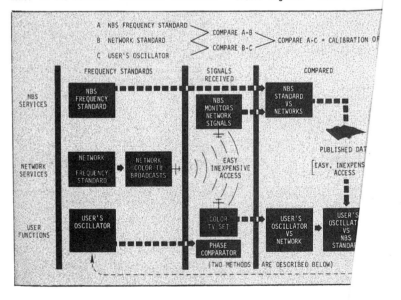

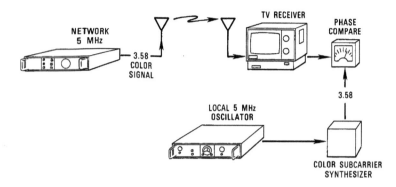

FIGURE 7.1. EQUIPMENT SETUP FOR COMPARING A 5 MHz OSCILLATOR WITH NETWORK 5 MHz OSCILLATOR.

bidium oscillators that produce the neces-
ry color signal at a frequency of 3.58 MHz.
1 home color television receivers "lock"
to the color subcarrier signal. So, when a
lor set is tuned to a network program, its
:ernal 3.58 MHz oscillator generates a rep-
·a of the atomic oscillator signal back at
network studio. This allows everyone with
olor set to have almost direct access to a
ber of atomic oscillators for calibrations.
not just any oscillators at that--these
checked by the National Bureau of Stan-
ls which publishes their exact frequency.

The 3.58 MHz signal from the color re-
er is not a substitute for your own oscil-
r. It is a calibrating signal that can be
to set your oscillator very accurately.
nly 15 minutes, you can get results that
d take hours or days to get using NBS
o stations.

Let's look at the basic principle of the
ce. The oscillators used by the networks
MHz units modified to include a synthe-
to generate a 3.5795454545...MHz (round-
3.58 MHz) color subcarrier signal. The
MHz was synthesized by taking

$$\frac{63}{88} \times 5 \text{ MHz}.$$

e wishes to adjust the frequency of a 5
scillator to agree with the network 5
t· can be done by utilizing the phase
or scheme shown in figure 7.1. You
ask,· why measure phase to calibrate
ncy? As shown by the meter in figure
vo frequencies are compared by making a
comparison. In this case, the signals
d are at 3.58 MHz, so the phase meter
ale deflection "reads" one cycle (360°)

of that frequency or, using the period of the
signal, about 279 nanoseconds full scale. If
the local oscillator frequency changes relative to the network color signal, the meter
moves. How much does it move? If those two
frequencies differed by one cycle per second,
the meter would deflect zero to full scale in
one second and then start over again in the
next second.

Notice, however, that if the oscillator
in the figure were of high quality, we could
expect a slow moving meter. Let's take an
example: Assume the crystal oscillator frequency is already set to within one part in
10^{10} of the network oscillator frequency. The
problem is to figure out how much the meter
will move; that is, how many degrees of phase
or nanoseconds will be accumulated in one
second. Set up the problem like this:

$$\frac{\text{Nanoseconds accumulated}}{1 \text{ second}} = 1 \times 10^{-10}.$$

The right side of the equation is a number
without dimensions (a numeric) so the left
side must match. This means that we multiply
the denominator (seconds) by 10^{+9} to get
nanoseconds for our units. This will cancel
the nanoseconds in the numerator, giving us a
numeric on the left side.

SOLVING: Number of nanoseconds accumulated in
10^{+9} nanoseconds (one second)

$$= 10^{-10} \times 10^{+9} = 10^{-1}$$

$$= \text{one-tenth of a nanosecond in one}$$
second.

This means that our meter will take ten seconds to move one nanosecond. Or, we can say

132

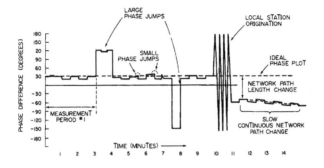

FIGURE 7.2. ILLUSTRATION OF PHASE INSTABILITIES INTRODUCED BY NETWORK PATH.

that our meter will move full scale (279 nanoseconds) in 2790 seconds. This is about 46 minutes.

So, measuring electrical phase is the way to go--it is really just another way of saying that we are measuring part of a cycle (called the relative frequency--Chapter 3), written:

$$- \frac{\Delta t}{T} = \frac{\Delta f}{f} ,$$

where the relative frequency, $\Delta f/f$, is our small fraction, 1×10^{-10}. Both methods described in the following pages are based on this principle of measuring phase to get frequency. The results for both methods give an answer as a relative frequency from the oscillator's nominal value.

If we synthesize 3.58 MHz from our 5 MHz oscillator, the phase of the two 3.58 MHz signals may be compared either on a meter or on a chart recorder. If the meter stands still or the chart recorder draws a straight line, indicating a fixed phase relationship between the two 3.58 signals, it follows that the two 5 MHz oscillators agree in frequency.

Of course, when we are using off-the-air TV signals, the meter will not stand still and the chart recorder won't draw a straight line. Due to instabilities in the propagation path, we will see small "bumps" or jumps in phase. In addition, the oscillator itself will affect the straightness of the line.

7.2.1 PHASE INSTABILITIES OF THE TV SIGNALS

If the 5 MHz local oscillator were perfectly matched with the network oscillator so that the only instabilities were those introduced by the network path, a phase chart

recording of the two signals might look something like figure 7.2.

There are basically four types of phase instability illustrated: large and small phase jumps, network path length changes, and local station originations. All of these instabilities tend to limit the precision of phase measurement, and hence, the precision with which you can calibrate the frequency of the local oscillator to the network 3.58 MHz. The ideal phase plot is shown as a straight dotted line at $+30°$, constant phase. The ultimate resolution is limited primarily by the slow continuous network path change illustrated from minutes 11 through 14. In most cases, resolution is limited to about 10 nanoseconds in 15 minutes. This corresponds to a resolution of:

$$- \frac{\Delta f}{f} = \frac{\Delta t}{T} = \frac{10 \text{ ns}}{15 \text{ min}} = \frac{10 \text{ ns}}{900 \text{ s}}$$

$$= \frac{10^{-8} \text{ s}}{0.9 \times 10^3 \text{ s}} = 1.1 \times 10^{-11}.$$

For measurement times shorter than 15 minutes, the measurement resolution will be reduced roughly in proportion to the reduction in measurement time; i.e., 1×10^{-10} in 1.5 minutes. Note that this resolution represents about the best one can hope to achieve. If large and small phase jumps are not taken into account, the result can be much worse.

The large phase jumps (up to $\pm 90°$ or ± 70 ns) are primarily the result of the way the networks operate. Phase jumps are caused by switching from one video tape machine or camera to another, with different lengths of cable being placed in the path. Most large phase jumps are caused by changes from a program to a commercial and back again. Using TV requires some attention to what is on the

133

screen--one should be alert to sudden phase jumps coincident with commercials.

Small phase jumps are the result of phase distortion in the microwave system used to carry the network programs and to multipath signals between the local station transmitter and the TV receiver. Differential phase distortion within the TV receiver also contributes. The magnitude of small phase jumps is on the order of 1 to 10 ns, depending on the network and the degree of multipath at the receiving location. For best results, you should use an antenna system that minimizes "ghosts."

During station breaks, the received 3.58 MHz often originates from the local television station's 3.58 MHz oscillator. Unless the local station is one of the few that is equipped with a rubidium or cesium oscillator, its frequency will probably be no better than 1×10^{-7}, and the phase will change more rapidly, about one full cycle per second. No precision measurements can be made on local programming. Many stations record network programs for rebroadcast at a different time. When network programs are "tape delayed," the 3.58 MHz, again, is referenced to the local station's oscillator and is therefore invalid as a precision reference. In any given area of the U.S., a few days of experience will provide a user with a good idea of local program schedules.

7.2.2 TYPICAL VALUES FOR THE U.S. NETWORKS

In all of the preceding discussions, it has been assumed that the goal is to calibrate a local oscillator and to make it agree exactly with the frequency of the network oscillators. This is not what we want. For reasons that need not concern us here, the frequency of the network oscillators differ from the NBS frequency standard by about -3000 parts in 10^{11}. The exact difference is measured by NBS and published in the monthly NBS Time and Frequency Bulletin. This publication can be obtained, free of charge, by writing the Time & Frequency Services Group, NBS, Boulder, CO 80303. In performing a calibration, the oscillator versus TV should give a difference equal to the published values.

The average relative frequency for the three commercial networks, with respect to the NBS frequency standard, for the week June 24-30, was:

EAST COAST

$$F_{NBC} = -3013.9 \times 10^{-11}$$

$$F_{CBS} = -3002.4 \times 10^{-11}$$

$$F_{ABC} = -3000.0 \times 10^{-11}$$

WEST COAST

$$F_{NBC} = -3015.2 \times 10^{-11}$$

$$F_{CBS} = \text{Not Monitored}$$

$$F_{ABC} = -3000.0 \times 10^{-11}$$

The minus sign preceding the results indicates that the network subcarrier signal is lower in frequency than the NBS frequency standard.

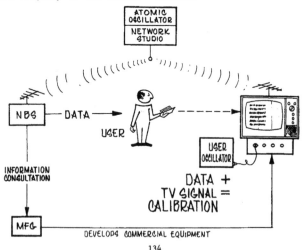

ATOMIC OSCILLATOR

NETWORK STUDIO

NBS — DATA →

USER

INFORMATION CONSULTATION

MFG

USER OSCILLATOR

DATA + TV SIGNAL = CALIBRATION

DEVELOPS COMMERCIAL EQUIPMENT

You might ask why there are two sets of data. Network programs in the Eastern, Central, and Mountain Time Zones originate from New York City. However, in the Pacific Time Zone, network programs originate from Los Angeles, so different oscillators are involved. The NBS monitors the networks on both coasts. Therefore, if you live in the Eastern, Central, or Mountain Time Zone, you should use the data published for the East Coast. The West Coast data are only for those users who live in the Pacific Time Zone.

To summarize our discussion thus far, a person who wants to calibrate an oscillator can use the television network color subcarrier to do so. This is possible because NBS checks the network frequencies and publishes the "offset" (or error) of the network oscillators with respect to the NBS standard. A user would know two things: (1) the difference between his oscillator and the network oscillator (by measurement), and (2) the difference between the network and NBS (by publication). With this information, he can compute the difference between his oscillator and NBS. Thus, his calibration is traceable to NBS.

So far, we have talked mostly about high-quality oscillators. TV signals are equally good for calibration of low-quality oscillators. In fact, without even looking at the NBS calibration data, you can depend on the color signal from any TV station to be in error by less than 3 parts per million, even if it is not showing a network program. The FCC requires this accuracy from all stations.

7.3 HOW RELATIVE FREQUENCY IS MEASURED

The following sections will discuss how instruments can be constructed to perform TV frequency calibrations. To have a basis for discussing actual instrumentation, two NBS-designed instruments are described. Commercial equipment is based on similar principles. For exact information on a particular instrument, the manufacturers' manuals should be consulted.

Two signals are involved in our measurement. One is the calibrated 3.58 MHz signal from the network via a color TV receiver. The other is a 3.58 MHz signal that we have generated from our own oscillator. These two signals will be phase compared and the output from the comparator used as an indicator to calibrate our oscillator. When you mix two signals and get their difference frequency, it is called a beat note. For a violin or piano, the beat note is often a few hertz. But for our oscillator it is much lower in frequency. In fact, it usually takes about 10 seconds for

one beat note to occur. Because of this, we simply time the period of the beat note to get the information we need to make a calibration. By measuring the time of the beat note period, we increase our measurement accuracy.

With some simple math, we can see how time measurements can be used for frequency calibrations. We will find the length (in time) of our signals, notice how fast they are moving, and then solve for the period of the beat note that results.

Now, for our 3.58 MHz signal, the period is:

$$\Delta t = \frac{1}{f} = \frac{1}{3.58 \text{ MHz}} = 27936 \times 10^{-11} \text{ seconds.}$$

This tells us how far either of our signals has to move in time or phase to get one cycle of the beat note. If we know how long it takes, we can solve for the offset. Recall that relative frequency is simply the numerical ratio of two frequencies. It can be obtained by the next equation (see Chapter 3).

$$\frac{\Delta f}{f} = - \frac{\Delta t}{T} = \frac{\text{HOW MUCH IT MOVES}}{\text{HOW LONG IT TAKES}}$$

$$= \frac{1 \text{ PERIOD OF THE 3.58 MHz}}{1 \text{ PERIOD OF BEAT NOTE}}$$

We know that $\Delta f/f$ for the U.S. TV networks is nominally -3000×10^{-11}, so we can rearrange the equation to see how long a beat note will take.

$$T = \frac{\Delta t}{- \left(\frac{\Delta f}{f} \right)} = \frac{27936 \times 10^{-11}}{3000 \times 10^{-11}}$$

$$= 9.31 \text{ SECONDS AS THE} \\ \text{PERIOD OF THE BEAT NOTE}$$

This means that if the relative frequency of the network color signal were exactly -3000×10^{-11}, as is the case with ABC, the network 3.58 signal would change one cycle with respect to an NBS-controlled 3.58 MHz oscillator in 9.31 seconds. We now have our time measurement scheme.

For example, if the network relative frequency were reported as -3010×10^{-11}, the period of the beat note would be shortened to:

135

$$T = \frac{27936 \times 10^{-11}}{3010 \times 10^{-11}} = 9.28 \text{ seconds.}$$

Let's look at this example more closely. A network frequency change of 10 parts in 10^{11} resulted in a beat note period change of $9.31 - 9.28 = 0.03$ second. So, if our timing measurement is in error by 0.03 second, the frequency measurement will be in error by 10 parts in 10^{11}. This tells us how well our time measurement scheme will work. An error of 0.03 second is not very likely with the electronic devices available today. Even stopwatches will be useful to us. We will now examine some methods of implementing the beat note period measurement as used in the equipment designed by NBS.

7.3.1 COLOR BAR COMPARATOR

One simple method devised by NBS to compare a local frequency source with the network oscillators is to use a color bar comparator. This is a small, low-cost circuit that can be added to any color TV receiver. It is shown in block form and schematically.

As shown on the block diagram, the circuit connects only to the TV receiver antenna terminals. However, better results can be obtained if the receiver is modified with an input connector that feeds the color bar signal directly to the color processing (Chroma) circuit in the set. This modification does not affect normal operation of the receiver.

A schematic of the NBS color bar comparator is included so that the circuit can be constructed and used. Notice the input jack J1 on the left. It accepts the local frequency source at a frequency of 10 MHz or at an integer submultiple; that is, 10 MHz divided by "N," where N = 1, 2, 3....up to 100. This lets you compare frequency sources of many different frequencies if you wish--one at a time, of course.

Two options are shown on the schematic: the antenna TV interface circuit and also a video TV interface circuit for one model of receiver. No matter how the signal gets into the receiver, it will be processed just as if it were normal picture information. The beat of the local crystal oscillator with the network signal forms a vertical "rainbow" bar. The color of the bar changes across the width of the bar as the oscillator being calibrated changes with respect to the network oscillator. The measurement consists of using a stopwatch to measure how long it takes this color change to occur. The stopwatch reading is equal to the period of the beat note.

To use the color bar system for oscillator calibration, tune the receiver to a network color program, and set the oscillator to be calibrated so that the rainbow appears to move across the bar from right to left through a complete color cycle (from red to green to blue and back to red) in about 10 seconds. The color bar itself moves slowly across the screen from right to left. If the frequency of your oscillator is far off, the colors in the rainbow pattern will change very rapidly and the entire bar will move rapidly across

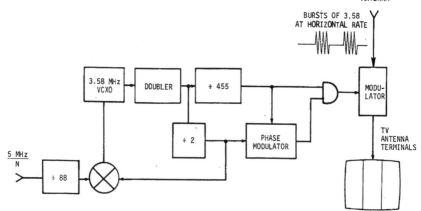

FIGURE 7.3.   BLOCK DIAGRAM OF COLOR BAR COMPARATOR.

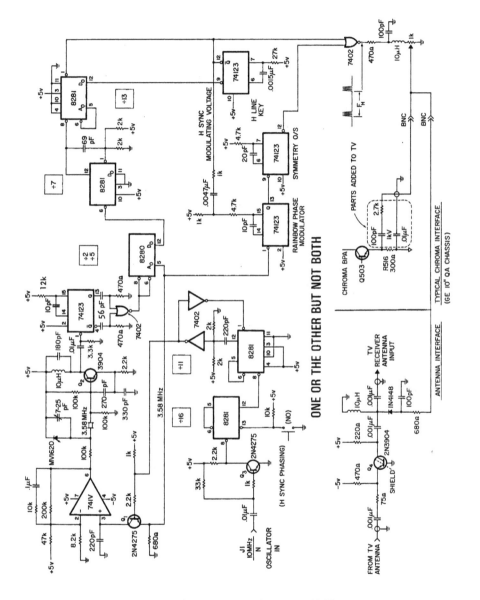

FIGURE 7.4. SCHEMATIC, COLOR BAR COMPARATOR

137

the screen in the direction of the color changes. The bar can be positioned to the middle of the screen by the pushbutton labeled "Horizontal Sync Phasing."

With the rainbow repeating colors in about 10 seconds, carefully adjust the crystal oscillator until the period is:

$$T = \frac{27936 \times 10^{-11}}{\text{NBS DATA FOR NETWORK BEING VIEWED}} \text{ seconds.}$$

If possible, check more than one network to increase the confidence in calibration. Also, by measuring over ten beat note periods, the effect of reaction time with the stopwatch will be reduced. Recall that for 1 period, an error of 0.03 second corresponds to a frequency error of 10 parts in 10^{11}. For ten periods, a measurement error of 0.03 second will result in a frequency error of 1 part in 10^{11}.

As mentioned before, the rainbow will be of higher quality if the signal is connected into the receiver as a video signal. The improvement is due to the fact that antenna injection also modulates the received audio carrier. This results in a visible beat note that constantly changes with the audio content of the program. The other example circuit shown connects the signal directly into the Chroma bandpass amplifier. The rainbow is then nearly perfect.

7.3.2 THE NBS SYSTEM 358 FREQUENCY MEASUREMENT COMPUTER

For those users who require accuracies approaching 1 part in 10^{11}, another calibration method is available. The system 358 Frequency Measurement Computer (FMC) computes and displays the oscillator's relative frequency directly on the TV screen. The user simply turns on the unit and waits for 15 minutes. He then comes back, reads off 10 values, and averages them to obtain the

result. Other versions of the FMC are commercially available that use a separate digital display rather than the TV screen.

The FMC automatically performs the operation of measuring the period of the 3.58 MHz beat note "T," computing $\Delta t/T$, scaling the result for readout in parts in 10^{11}, and displaying the one- and ten-period averages on the TV screen. Readout is a series of 4-digit numbers, representing the relative frequency between the oscillator being calibrated and the atomically-controlled color subcarrier. By referring to network data published monthly in the NBS Time and Frequency Bulletin, a user of the FMC can set his oscillator to agree with the NBS standard to better than 1 part in 10^{10} in five minutes and 3 parts in 10^{11} in fifteen minutes. Radio methods of calibrating oscillators traceable to NBS require several days of "averaging" to achieve this resolution. Measurement uncertainties as a function of averaging time that may reasonably be expected are:

AVERAGING TIME	MEASUREMENT UNCERTAINTY
10 seconds	3×10^{-10}
100 seconds	6×10^{-11}
15 minutes	2×10^{-11}
30 minutes	1.5×10^{-11}

When making network frequency comparisons, the FMC will also accept as input any oscillator whose frequency is 10/N MHz, where "N" is any integer from 1 to 100. The highest available frequency should be used when there is a choice.

The FMC consists of a 5-inch color TV receiver with an additional electronics package mounted to its base. Readout for the FMC is on the TV screen, with two columns of ten 4-digit numbers, and an additional analog "phase cursor."

A. Block Diagram Overview

The FMC is divided into four main functional parts: relative frequency scaler; beat note phase comparator and cursor generator; data store and display; and relative frequency generator. In this section, the term "offset" is used interchangeably with "relative frequency."

1. Relative Frequency Scaler. The relative frequency or offset scaler processes the beat note and 3.58 MHz signals to generate an output count proportional to relative frequency x 10^{-11}. This output count is applied directly to the single-period 4-digit counter.

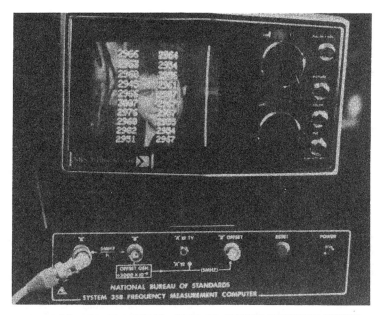

FIGURE 7.5. THE NBS PROTOTYPE, SYSTEM 358 FREQUENCY MEASUREMENT COMPUTER.

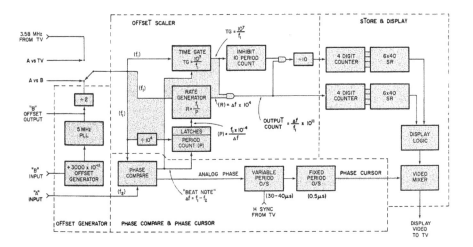

FIGURE 7.6. BLOCK DIAGRAM, SYSTEM 358 FREQUENCY MEASUREMENT COMPUTER.

139

A new measurement is stored in the 6 x 40 single-period shift register (SR) for readout. The output count of the scaler is also applied to the 4-digit ten-period counter through an inhibit gate and a ÷ 10 prescaler. The ten-period counter therefore accumulates 1/10 of its total count on each single-period measurement.

Theory of the Scaler. The scaler accepts the beat note and the color subcarrier frequency, f_1, from the TV. The variable divider generates output pulses at the rate R = Δf x 10^4; that is, 10,000 times the beat note frequency. These rate pulses are gated on for a time equal to $10^7/f_1$. The count output to the data store counters is therefore:

$$\frac{\Delta f}{f} \times 10^{11} \quad (\text{count} = R \times TG = \Delta f \times 10^4 \times \frac{10^7}{f_1}).$$

The objective is to obtain a reading on a 4-digit counter at the output such that each accumulated count is equal to 1 part in 10^{11} relative frequency between f_1 and f_2. We can verify that this has been accomplished by the following analysis. First, relative frequency is defined as:

$$\text{RELATIVE FREQUENCY} = \text{OFFSET} = \frac{f_1 - f_2}{f_1} = \frac{\Delta f}{f_1}.$$

The beat note comparator generates $f_1 - f_2 = \Delta f$. The beat note frequency, Δf, is applied as a start-stop gate to the period counter, "P." The total count accumulated in "P" in one cycle of Δf can be expressed as:

$$"P" = \frac{f_1 \times 10^{-4}}{\Delta f}.$$

For each cycle of Δf, the period count is latched and applied to the rate generator, where the output rate is

$$R = \frac{f_1}{P}.$$

The rate of output pulses from divider (÷P) is:

$$R = \frac{f_1}{P} = \frac{f_1}{\dfrac{f_1 \times 10^{-4}}{\Delta f}} = \frac{\Delta f}{10^{-4}} = \Delta f \times 10^4.$$

This rate output is accumulated in the 4-digit counter for a period of time determined by time gate, TG, which enables the "and" gate preceding the counter. The total output count for each measurement cycle can be expressed as:

$$\text{OUTPUT COUNT} = R \times TG$$

$$= \Delta f \times 10^4 \times \frac{10^7}{f_1} = \frac{\Delta f}{f_1} \times 10^{11}.$$

Therefore, if $\Delta f/f_1 = 3000 \times 10^{-11}$, the output count accumulated will be

$$\frac{\Delta f}{f_1} \times 10^{11} = [3000 \times 10^{-11}] \times 10^{11}$$

$$= 3000 \times 10^0 = 3000 \text{ counts}.$$

The scaling factor for the output is the product of the 10^4 divider and 10^7 time gate. These two factors may be partitioned in other ways--for example, 10^5 and 10^6 or 10^3 and 10^8; however, the partitioning used here is optimum for the nominal 3000 parts in 10^{11} offset to be measured.

With the partioning used and an offset of 3000 parts in 10^{11}, the period count will be 3,333 and the time gate, TG, will be on for 0.3 of the beat note period. The maximum offset that can be measured is 10,000 parts in 10^{11}, in which case the time gate is on continuously, and simultaneously, the 4-digit counter overflows. The minimum offset that can be measured is determined by overflow of the 16-bit period counter, P. This occurs at an offset of approximately 150 parts in 10^{11}.

Neither the 3.58 MHz color subcarrier nor the 10/N MHz input to be compared with it has been mentioned in the preceding discussion. Normally, f_1 is the 3.58 MHz input and f_2 is the harmonically related 10/N MHz input. However, the unit will work equally well with any frequencies that have harmonically related offsets in the range of 150 to 10,000 parts in 10^{11}.

2. Beat Note Phase Comparator and Phase Cursor Generator. The phase comparator section compares the phase of each 126th cycle of 3.58 MHz from the TV receiver with each 176th cycle of the "A" input. This circuit performs the same harmonic synthesis function as the network oscillators which generate 3.579545454 ... MHz by taking 63/88 x 5 MHz. The analog

voltage resulting from this phase comparison is stored in a sample-and-hold integrator. The analog "beat note" is further processed through two paths, the phase cursor path for a visual indication of the beat note and a Schmitt trigger to condition the signal for the TTL offset scaler input.

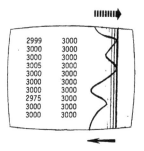

3. **Data Store and Display.** The gated frequency from the offset scler is accumulated in the two 4-digit counters in the data store-and-display section. After each beat note measurement cycle, the single-period counter is gated on for 2.79 seconds. The accumulated single-period count is then dumped to the single-period store for readout. The 10-period 4-digit counter is preceded by a ÷10 so on each single-period average, it accumulates 1/10 of its total count. At the end of ten 1-period averages, the 10-period counter contents are dumped to the 10-period store for readout.

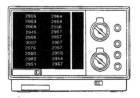

The readout data are presented as two columns of ten 4-digit numbers. The left column represents single-period offset readouts and the right column represents ten-period offset readouts. To start a measurement sequence, the user pushes the reset button. All readouts are reset to zero and the top 4 digits in each column are intensified, indicating that data will be loaded in these positions.

At the end of approximately 13 seconds, the first single-period measurement is com-

pleted and the data are loaded into the top 4 digits in the left column. The second 4 digits in the left column will now be intensified. On each following 10-second interval, data are loaded into succeeding positions in the left column until ten single-period averages have been accumulated. At this time, the 10 single-period measurements are averaged and then loaded into the top 4 digits of the right-hand column. This process continues until all 10-period averages have been loaded.

Two operator aids are included in this NBS equipment. If a large phase jump occurs at any time in the sequence, the corresponding single-period offset will be read out, but that measurement will not be incorporated into the 10-period average. On the next measurement cycle, that single-period average will be rewritten and the count incremented. The TV speaker "beeps" as a reminder to the operator each time a measurement is completed.

B. **Use of FMC with Unstable Crystal Oscillators**

The FMC is primarily intended to calibrate with stability of better than 1 part in 10^7 per day. In fact, if the oscillator being calibrated will not hold still to better than 1 part in 10^7 (± 10,000 parts in 10^{11}), the digital readout cannot be used. About the best one can do with a really poor-quality oscillator is make the cursor "stand still" momentarily.

If the cursor moves less than one cycle each three seconds, then the oscillator is within 1 part in 10^7 of the color subcarrier. By switching between networks, one can obtain a "concensus;" however, there is no way to really verify if the subcarrier is originating from a network, as can be done by checking a good oscillator.

The FCC requires all stations to maintain 3.5795454545...MHz ± 10 Hz on their subcarriers, so by comparing with several stations, one is assured of being within 3 parts in 10^6. If the oscillator being calibrated is stable to 0.001% (1 part in 10^5), it is close enough.

7.4 **GETTING GOOD TV CALIBRATIONS**

Although the frequency calibration service using TV reduces the labor of calibrating an oscillator, it does not relieve the user of the responsibility of verifying that "good" data are actually being processed. The user must be sure that the station being used is actually transmitting network-originated programs. (See Section 7.5).

141

At the present time, data for three networks are available from NBS. The easiest thing to do is to take successive readings, if possible, on all three networks for 1.5 minutes each and verify that they differ relative to each other by approximately the correct amount. Nominally, ABC operates at exactly -3000 parts in 10^{11}.

Assume relative frequency readings were taken with the following correct results:

$$F_{ABC} = -3131 \times 10^{-11}$$
$$F_{NBC} = -3150 \times 10^{-11}$$
$$F_{CBS} = -3130 \times 10^{-11}$$

The readings are consistent from one network to another. If one of the three had differed radically from its correct relative value, we could assume that the station involved was on local programming. For example, if the reading for NBC were -2460, then the NBC station would not be used at that time.

Measurements must be made only when network-originated programs are on the air. Network stations in the Eastern, Central, and Pacific time zones typically carry 10 to 14 hours of network-originated programs per day. The Mountain time zone carries about four hours per day.

It should be noted that network-originated programs are not necessarily "live" programs (such as football or basketball games). It simply means that the programs are originated from the network centers in either Los Angeles or New York City and are not tape delayed by the local stations. Programs originating at the local stations (such as the channel 7 news as opposed to CBS Evening News) cannot be used.

The best choices are the daytime soap operas and quiz shows. These usually are daily programs and originate from the New York and/or Los Angeles network centers. Weekly evening programs are usually network-originated except in the Mountain Time Zone. Once the programming schedule in any given city is identified, a log can be made to indicate when measurements can be made. There is, however, one further word of caution (see below).

7.5 THE DIGITAL FRAME SYNCHRONIZER

Some local stations are now using digital frame synchronizers on their incoming network lines. This is true of the ABC stations in the San Francisco and Chicago areas, which provide network feeds to other local stations in the area, as well as many other local stations throughout the country.

The frame synchronizer stores one complete TV frame (1/30 second) in digital memory and "reads" the TV picture under control of the local station sync generator. Since most sync generators are driven by a crystal oscillator, the color subcarrier is no longer referenced to the network atomic standard, and the signals cannot be used for frequency calibration.

The ABC stations in Los Angeles and New York also use frame synchronizers on the incoming network lines, but these stations are co-located with the network studios and use the network cesium standards for their 3.58 MHz reference. Therefore, WABC and KABC still provide stable frequency references and can be used for frequency transfer measurements.

In 1978, a questionnaire was sent to all network-affiliated TV stations to determine which ones are using or plan to use frame synchronizers. The results of this survey are available upon request from the Time and Frequency Services Group, National Bureau of Standards, Boulder, CO 80303.

At present, only a few of the largest TV stations have two or more frame synchronizers. The smaller stations that have a single frame synchronizer will use it for their "electronic news," not to hang across the incoming network line. Of course, this situation may change in five years if the price of synchronizers drops.

If you are using the TV networks for frequency transfer, the best precaution is simply to use more than one network so that data can be verified.

7.6 USNO TV FREQUENCY CALIBRATION SERVICE

The USNO also offers a TV frequency calibration service to users who can receive stations WTTG (Washington, D.C.), WAPB (Annapolis, MD), or KTTV (Los Angeles, CA). The USNO has stabilized the frequency of the cesium standards at these stations, so the service is somewhat different than the NBS service. For information, contact the Time Services Division, U. S. Naval Observatory, Washington, DC 20390.

WORSHEET FOR USE WITH COLOR BAR COMPARATOR			
DATE:		START TIME:	
NETWORK	NBC	CBS	ABC
ENTER PUBLISHED DATA			
COMPUTE PERIOD*			
STOPWATCH TIMES			
AVERAGE:			
COMPUTE RELATIVE FREQUENCY**			

*

$$\frac{27936 \times 10^{-11}}{\text{PUBLISHED DATA}} = \frac{27936 \times 10^{-11}}{\text{XXXXXX}} = \text{XXXX SECONDS BEAT NOTE PERIOD.}$$

**

$$\frac{27936 \times 10^{-11}}{\text{MEASURED TIME}} = \frac{27936 \times 10^{-11}}{\text{XXXXXX}} = \text{XXXXXX RELATIVE FREQUENCY OF YOUR OSCILLATOR.}$$

Compare PUBLISHED DATA with COMPUTED RELATIVE FREQUENCY to determine difference between your oscillator and the NBS frequency standard.

WORKSHEET FOR USE WITH DIGITAL OFFSET COMPUTER			
DATE:		START TIME:	
NETWORK	NBC	CBS	ABC
ENTER PUBLISHED DATA			
10-PERIOD AVERAGES			
100-PERIOD AVERAGE			
YOUR OSC. VS NBS			

Compare the 100-period average with PUBLISHED DATA for each network. If it is higher than PUBLISHED DATA, then the oscillator being calibrated is too high in frequency. If it is lower, then the oscillator is too low in frequency.

FIGURE 7.7. SAMPLE WORKSHEETS FOR USING TELEVISION SUBCARRIER FOR FREQUENCY TRANSFER.

The previous chapter discusses the use of the TV network color subcarrier for frequency calibrations. The television synchronization pulses can also be used as a transfer standard for both time and frequency.

A television picture is made up of 525 lines which are scanned or traced to produce the image you see on the screen. First, the odd lines (called the odd field) are traced, then the even lines (the even field) are traced in between the odd. This scanning happens so fast (60 times a second) that your eyes cannot detect it and you have the illusion of seeing a continuous picture.

Line-10 (odd) is one of the 525 lines that make up each picture; it can be used for both frequency and time calibrations. It was chosen because it is easy to pick out from the rest of the lines with relatively simple electronic circuits. To calibrate an oscillator, a clock pulse from the oscillator is compared to the trailing edge of the Line-10 (odd) synchronizing pulse. Measurements by users of this service, made with specially-designed equipment, will correspond to measurements made at NBS at the same time.

Now that we know what Line-10 is, let's say what it isn't. It isn't the color subcarrier! In the previous chapter, we discussed using the color subcarrier as a frequency calibration source. Line-10 has nothing to do with the color subcarrier system. Line-10 works with either color or black and white television sets. Some equipment that is designed for the color subcarrier system will have a Line-10 output pulse because it is handy to do so, but the techniques involved are entirely different for the two TV-based systems. They do have one thing in common, though. All measurements using either system must be made on network (major U.S. television networks) programming if NBS traceability is to be achieved. In addition, Line-10 measurements must be made at exactly the same time of day that NBS makes its measurements. (Neither of these requirements apply if you are using Line-10 on a local basis.) This will be discussed more fully later on.

FIRST THE ODD FIELD IS TRACED.

THEN THE EVEN FIELD IS TRACED.

FIELDS COMBINE TO FORM
COMPLETE PICTURE.

8.1 HOW THE SERVICE WORKS

Assuming that you have a TV receiver and a Line-10 pulse is available, what do you do with it? You can use it to set your clock or to measure and/or set the frequency of your oscillator. It provides a means of checking oscillators to within parts in 10^{11}, but the oscillator/clock should be of fairly high quality (1 part in 10^8 or better).

8.1.1 USING LINE-10 ON A LOCAL BASIS

To understand how the service works, let's discuss a real-life example: The Line-10 method is used routinely to compare the NBS frequency standard in Boulder, Colorado, with the clocks at radio stations WWV and WWVB in Ft. Collins, Colorado. Both locations are within common view of the Denver television stations. Let's assume that the clocks in Boulder and Ft. Collins are synchronized. Now suppose we measure the arrival time of the TV Line-10 synchronization pulse at Boulder (location A) with respect to the NBS standard. A short time later, this same pulse will arrive at Ft. Collins (location B) and we can measure its arrival time with respect to the Ft. Collins clock. Since the clocks are synchronized, the difference in the two clock readings is only the differential propagation delay, written $t_B - t_A$ as shown in figure 8.1.

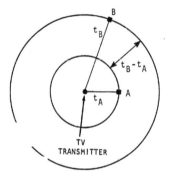

FIGURE 8.1 DIFFERENTIAL TIME
DELAY, $T_B - T_A$.

We now know the propagation delay between Boulder and Ft. Collins. If we make another measurement and find that the difference in the two clock readings is not equal to the propagation delay, then this means that one of the clocks is running faster or slower than the other. The discrepancy represents the time difference between the clock at Ft. Collins and the clock at Boulder.

The same technique can be used by anyone who wants to transfer time and/or frequency from one location to another. This works only when both locations can see the same TV signal. An example of using this technique would be a factory with scattered branches that wants to calibrate an oscillator at one of the branches that is not within cabling distance of the main factory. By using two Line-10 units, the factory and the branch can both look at a local TV signal and effect the transfer.

Even if you don't know the propagation delay, the Line-10 technique can be used to determine the frequency stability of your oscillator compared to another. Using a divider, simply operate your oscillator as a clock. By watching the day-to-day differences, you can tell if your clock is gaining or losing with respect to the clock used for comparison.

Note that such a local signal can be any kind of TV signal, not necessarily network TV. Also, the accuracy of transfer using this method can be very great. The NBS link to Ft. Collins typically can achieve ten nanoseconds time transfer. Local use of Line-10 is a cheap way to distribute frequency and time at high accuracy levels.

8.1.2 USING LINE-10 FOR NBS TRACEABILITY

In figure 8.1, we assumed that both locations were within common view of the same television station. However, in actual practice, a given TV program can be observed simultaneously over a large geographical region because of the TV microwave distribution network. For example, this technique has been used to compare clocks in the Boulder and Washington, D. C. areas by making simultaneous measurements on programs originating from New York City (see figure 8.2).

NBS routinely makes readings on the three major networks on both the East and West Coasts. These data are published monthly in the NBS Time and Frequency Bulletin, which is available upon request. A sample page of the Bulletin is shown in figure 8.3.

If you want to compare your clock with the NBS standard, you must make your measurements at the same time as NBS and, in addition, the measurements must be on programs

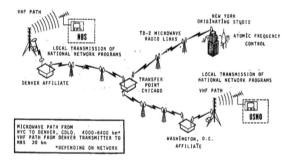

FIGURE 8.2 TYPICAL ROUTING OF TV SIGNALS FROM NEW YORK CITY
ORIGINATING STUDIOS TO DISTANT RECEIVERS.

146

EAST COAST

AUG.	NBC		CBS		ABC	
	19:25:00	19:31:00	19:26:00	19:32:00	19:27:00	19:33:00
1						
2	15694.75	08672.31	31093.91	24071.52	15667.06	08644.55
3	32005.47	24983.03	14018.02	06995.50	31872.57	24850.06
4	14949.71	07927.26	30308.97	23286.38	14884.43	07861.91
5	31260.82	24238.38	13233.12	06210.59	31090.14	24067.63
6	14205.24	07182.67	29523.82	22501.28	14015.66	06993.12
7						
8						
9	29771.69	22749.24	11663.10	04640.58	29525.03	22502.51
10	12715.81	05693.36	27954.14	20931.62	12450.44	05427.93
11	29026.94	22004.47	10878.29	03855.73	28742.49	21719.93
12	12058.43	05035.96	27168.97	20146.43	11667.93	04645.40
13	28282.51	21260.09	10093.04	03070.54	27960.03	20937.53
14						
15						
16	00259.24	26603.42	25598.64	18616.28	15259.36	08236.83
17	16567.44	09542.88	08522.81	01500.38	31551.49	24529.06
18	32873.49	25851.00	24813.56	17791.00	14477.22	07454.66
19	15814.94	08792.46	07737.72	00715.15	--	23746.32
20	32123.48	25101.04	24029.13	17006.61	--	06699.40

FIGURE 8.3.   NBS LINE-10 DATA FROM NBS TIME & FREQUENCY BULLETIN.

originating from a common source--New York City (for the Eastern, Central, and Mountain time zones) or Los Angeles (for the Pacific Time Zone). Currently, measurements are made by NBS on the East Coast networks at the following times: NBC--2025 and 2031 UTC; CBS--2026 and 2032 UTC; ABC--2027 and 2033 UTC (During Daylight Saving Time, the measurements are made beginning at 1925 UTC.) West Coast measurements (furnished by the Hewlett-Packard Company in Santa Clara, California) are made at: NBC--2125 and 2131 UTC; CBS--2126 and 2132 UTC; ABC--2127 and 2133 UTC (all year round).

The measurements at NBS are made on the trailing edge of the horizontal Line-10 (odd) sync pulse which occurs approximately every 1/30th of a second. This means you must already know your time to 1/30 of a second or better to be assured that you are measuring the same pulse that is being measured by NBS.

All of this sounds fairly straightforward, but it isn't all that tidy. First of all, the calibration isn't available to the user until he receives the data he needs in the mail. This means that he must keep careful Line-10 records and wait for the results.

Meanwhile the clock must be kept running so that when the data arrive and the comparison is made, the clock error can be determined and something done about it. This means that the clock involved must be a pretty good one. It should be as good as 1 part in 10^8 or better. Otherwise, the whole operation will be labor-intensive and the user will not be able to predict the time very well. Nor will he know the frequency very accurately if the system has large variations. Line-10 provides a means of checking oscillators to within parts in 10^{11}, but the oscillators must be of high enough quality to deserve the investment of time and effort. If you need lower accuracy, there are cheaper ways that require less effort.

8.2 EQUIPMENT NEEDED

To use the Line-10 technique, you need a black and white or color TV receiver, a line-10 synchronized pulse generator, a digital counter-printer, and a precision clock. Line-10 equipment is commercially available. The TV receiver is adapted to bring out the composite video signal that connects to the

147

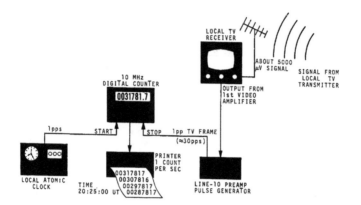

FIGURE 8.4 EQUIPMENT NEEDED FOR TV LINE-10 METHOD OF CLOCK SYNCHRONIZATION.

Line-10 pulse generator. This pulse is compared on a time-interval counter with a local one-pulse-per-second clock. The display on the counter (start with local clock, stop with Line-10) will be a series of numbers (see figure 8.4).

8.3 TV LINE-10 DATA, WHAT DO THE NUMBERS MEAN?

How do we get the numbers and what do they mean? First of all, let's review the basis for the frequencies used by commercial television systems in the U. S. The FCC specifies the frequency used for the chrominance subcarrier to be 3.5795454 ... MHz ± 10 Hz. The networks get this number by taking a 5 MHz oscillator and multiplying its output by 63 and dividing by 88. Then, the FCC states that the horizontal rate for the picture shall be 2/455 of the color subcarrier. This is 15.7342657 kHz. Next, you get the vertical rate by taking 2/525th of the nominal 15.7 kHz, giving a vertical rate of 59.94 Hz. But Line-10, the one we want, only occurs every other field so its rate is half of the vertical rate, or 29.97 Hz, and has a period of about 33 milliseconds--33.3666666... to be exact.

Now, in our Line-10 instrumentation set-up, we start our time interval counter with our clock at a 1 pps rate and stop it with the next Line-10 pulse. A stream of Line-10 data is shown in figure 8.5.

A reading occurs each second, each reading being slightly different from the one preceding it. Let's look at that difference. Notice the second and third columns which are

reading milliseconds. The number increases progressively by 1 millisecond per second. This is because Line-10 occurs at a frequency rate that is not an integer and appears to drift relative to our 1 pps. When the count approaches 33 milliseconds total, the time interval counter is stopped by the next Line-10 pulse that comes along, but it must always occur before 33 milliseconds elapse.

So instead of reading the number above the double line plus the 1 millisecond per second which would equal 33.3101 + 1.00000 = 34.3101, it must stop at the very next Line-10 pulse which appeared at the counter stop input at 34.3101 - 33.3666 = 0.9435. This is what

NBC LINE-10 (EAST COAST) FEBRUARY 11		
UTC	ms	μs
20:35:41	025	309.93
20:35:42	026	309.96
20:35:43	027	309.99
20:35:44	028	310.02
20:35:45	029	310.05
20:35:46	030	310.09
	031	310.11
	032	310.14
	033	310.17
	000	943.53
	001	943.56
	002	943.59
	003	943.62
	004	943.65

FIGURE 8.5. PRINTOUT OF NBC LINE-10 DATA.

148

| MEASUREMENT PERIOD | AVERAGE RELATIVE FREQUENCY (in parts in 10^{11}) | | | | | |
| | EAST COAST | | | WEST COAST | | |
	NBC	CBS	ABC	NBC	CBS	ABC
2 - 6 FEBRUARY	-3019.4	-2963.0	-3012.3	-3019.8	-	-3015.2
9 - 13 FEBRUARY	-3019.5	-2963.1	-3012.4	-3019.8	-	-3015.4
16 - 20 FEBRUARY	-3019.6	-2963.2	-3012.7	-3019.8	-	-3015.7

FIGURE 8.6. TELEVISION NETWORK FREQUENCIES RELATIVE TO THE NBS FREQUENCY STANDARD AS PUBLISHED IN THE NBS TIME & FREQUENCY BULLETIN.

was printed out. The time interval counter cannot ever print a number bigger than the period of the highest frequency on the "stop" input. Notice, too, that the last column in the data is also changing. The counter-printer displays 0.01 microsecond in the last digit and is changing at a rate of 30 nanoseconds (0.03 microsecond) per second.

How does this happen? In addition to operating at a frequency that is a non-integer, each TV station will have an error because its oscillator frequency differs from the NBS frequency standard. It differs enough to show up on a high resolution record like this one. It shows up on the published NBS data in figure 8.6. On the date these data were recorded, NBC's relative frequency was -3019.4 parts in 10^{11}; that is, just over 3 parts in 10^8. Just like any clock that is running at a different rate, it will accumulate a time error. In this case, the time error due to the basic frequency difference between NBS and NBC is 30.194 nanoseconds per second. So our data will show this amount of change, and it does. Since NBC's frequency is low relative to NBS, and its Line-10 pulse is stopping the clock (our time interval counter, in this case), the Line-10 pulses arrive later and later by the 30.194 nanosecond amount each second. This is evident in the last column of the printout (figure 8.5) as an increasing number since the time interval counter counts just a little longer each second. If local programming is used, such as between Boulder and Ft. Collins, this number can change by as much as 50 or 60 nanoseconds per second. It can also decrease rather than increase.

Although you can print out several numbers for each network, only the first number is used--the one that occurs right on the exact second that NBS is making measurements. Taking several readings insures that the equipment is working properly. Also, if you happen to miss the first second, you can

reconstruct the data. This is not a recommended practice, but it can be done.

8.4 WHAT DO YOU DO WITH THE DATA?

Assume that data are taken daily, careful records are kept, and the clock-oscillator combination being used for TV Line-10 data is not disturbed. After all, if you are recording data for 30 days and waiting for a check with NBS, you must not alter the setup. The data points can be tabulated or, better yet, plotted on graph paper. The accumulated time error of the oscillator is a measure of its frequency offset. Therefore, the data serve to calibrate the relative frequency of the oscillator with respect to NBS. Notice that we are making and recording time measurements to get frequency. As you will read later, we can also get time, but we must make some further measurements to get it.

Figure 8.7 shows the raw data for an oscillator being calibrated with the TV Line-10 system. The counter used for this measurement had a resolution of 0.1 microsecond. For the data taken, it appears that the oscillator changed about 10 microseconds a day relative to NBS. This works out to be about 1 part in 10^{10} frequency error. Having three major U.S. networks gives a good self-check on the system accuracy. Although the data will be slightly different for each of the three networks, the calibration error will be small.

8.5 RESOLUTION OF THE SYSTEM

How good is the system? It has been known to produce consistent results of around 1 part in 10^{11} over a period of several years. If you are willing to pay attention to details, you can achieve similar results. Of course, you must have a fairly good TV signal.

149

NBS LINE-10 DATA* JUNE		USER'S LOCAL CLOCK LINE-10 DATA	DIFFERENCE** NBS -	DAILY ACCUMULATION	TOTAL ACCUMULATION
DAY	(NBC)	(NBC)	LOCAL CLOCK	(MICROSECONDS)	(MICROSECONDS)
7	02216.54	05339.9	-3123.4	0	0
8	19526.34	22660.1	-3133.8	-10.4	-10.4
9	02374.93	05518.6	-3143.7	- 9.9	-20.3
10	18779.21	21934.0	-3154.8	-11.1	-31.4
11	01722.58	04888.5	-3165.9	-11.1	-42.5

*Published in NBS Time & Frequency Bulletin.

**This heading establishes a method for the user to learn if his clock is gaining or losing time. Since the NBS clock is always "on time," the user's data are subtracted from the NBS reading. It is as if a time interval counter produced this column where the NBS clock opens the gate and the user's clock closes it.

COMPUTING THE OSCILLATOR FREQUENCY

(Refer to Chapter 3 for explanation of the math)

μs ACCUMULATED IN 5 DAYS: -42.5 $\qquad \dfrac{\text{NO. OF MICROSECONDS (AVERAGE)}}{\text{1 DAY}} = \dfrac{10.6}{T} = \dfrac{\Delta t}{T}$

But one day has $86,400 \times 10^6$ microseconds. So:

$$-\frac{\Delta t}{T} = \frac{-10.6}{86,400 \times 10^6} = -1.23 \times 10^{-10} .$$

Also: $\qquad -\dfrac{\Delta t}{T} = \dfrac{\Delta f}{f} = 1.23 \times 10^{-10}.$

This is the relative frequency of the user's oscillator with respect to NBS. The actual frequency of the user's oscillator can also be found. If we assume this is a 2.5 MHz oscillator, then:

$$\frac{\Delta f}{f} = \frac{f_{actual} - f_{nominal}}{f_{nominal}} .$$

Substituting $\qquad 1.23 \times 10^{-10} = \dfrac{f_{actual} - 2.5 \times 10^6}{2.5 \times 10^6} .$

Solving: $\qquad f_{actual} = 2,500,000.0003075 \text{ Hz.}$

So the user's oscillator is higher than it's nominal value by a very small amount.

FIGURE 8.7. COMPUTATION OF OSCILLATOR FREQUENCY USING TV LINE-10 DATA.

It also helps your results if you can receive all three networks and cross-check the results for consistency. On the other end of the spectrum, would it be worthwhile to use the Line-10 system for lower-quality oscillators? The answer is yes, but up to a limit. Aside from the fact that other frequency calibration services are available on radio, the basis for a decision of whether to use Line-10 or not rests with the amount of time the user is willing to invest. For, as the oscillator quality goes down, the amount of attention the system requires increases. It becomes necessary to look at the data more closely and recognize and calibrate the larger frequency shift that will occur.

If the user's oscillator has a frequency error of 1 part in 10^8, it will gain or lose about 1 millisecond per day. Since the TV Line-10 pulse is ambiguous (it occurs about every 33 milliseconds), it would not be too hard to keep track of the data. However, if the local oscillator drifted and was off as much as 1 part in 10^6, the data would get harder to interpret. This would allow a change of 100 milliseconds per day and a different Line-10 pulse would be viewed by the user than that seen and reported by NBS. This is the case for any clock that has a large time error. Then, the user is stuck with either noting the differences and plotting them or correcting the data for whole periods of the Line-10 sync signals. This is 33.3666 ... milliseconds. So, although just about any clock oscillator can be used, it is not as simple and straightforward if large clock drifts are involved.

8.6 GETTING TIME OF DAY FROM LINE-10

Since the Line-10 data are directly traceable to NBS, it contains time-of-day information. As noted in the above paragraph, it is ambiguous to 33 milliseconds. This means that if you want to set a clock accurately (to within microseconds), you must first set it to within 33 milliseconds using some other scheme. A good way would be to use radio station WWV and an oscilloscope. Once you are within the ambiguity range of the Line-10 data, you can use Line-10 to set the clock more precisely. You would need to know the path and equipment delays. The path delay would just about have to be measured by carrying a clock to the NBS in Boulder. Once the delays are known, and if the user records data regularly from several networks, chances are he can keep time to within several microseconds using Line-10.

The need to calibrate the path and resolve the ambiguity may seem to penalize this system and suggest to the user that he use some other method, but there are not very many ways to get accurate timing. Chapter 9 describes Loran-C techniques for accurate time setting. It, too, needs to be calibrated and is ambiguous in the same way. So the steps are the same and Line-10 costs less.

8.7 THE DIGITAL FRAME SYNCHRONIZER

As discussed in Chapter 7 (page 142), some local TV stations are using digital frame synchronizers which affect the use of these stations for frequency calibrations. The frame synchronizer also eliminates the use of these stations for Line-10 time transfer when it is in the differential path. The ABC stations in Los Angeles and New York also use frame synchronizers on the incoming network lines so Line-10 reference is lost. For more information on the stations that are using frame synchronizers, contact the Time and Frequency Services Group, National Bureau of Standards, Boulder, Colorado 80303.

8.8 MEASUREMENTS COMPARED TO THE USNO

The USNO also publishes Line-10 measurements on the East Coast TV networks originating in New York City. Their measurements are made at the same time as the NBS measurements. Users desiring to compare their clocks with the USNO may obtain the USNO data from their Time Services Bulletin, Series 4. For more information, write to the Time Services Division, U. S. Naval Observatory, Washington, D. C. 20390.

8.9 LINE-10 EQUIPMENT AVAILABILITY

There are several manufacturers of TV Line-10 equipment. The NBS has also designed equipment to recover the Line-10 pulse from a commercial TV receiver. The circuit diagram is shown in figure 8.8. The instrument accepts 0.5 to 1.5 volts of video signal of either polarity from a TV receiver and makes available a Line-10 pulse referenced to either the leading or trailing edge of sync. (Note that the Line-10 measurements published in the NBS Time and Frequency Bulletin are referenced to the trailing edge of the Line-10 sync pulse.) The problem is that you will have an error equal to the width of the sync pulse if you use the leading edge and NBS uses the trailing edge. This design uses standard TTL integrated circuits and could be assembled by a skilled technician.

For more information about manufacturers of Line-10 equipment, write to the Time and Frequency Services Group, National Bureau of Standards, Boulder, CO 80303.

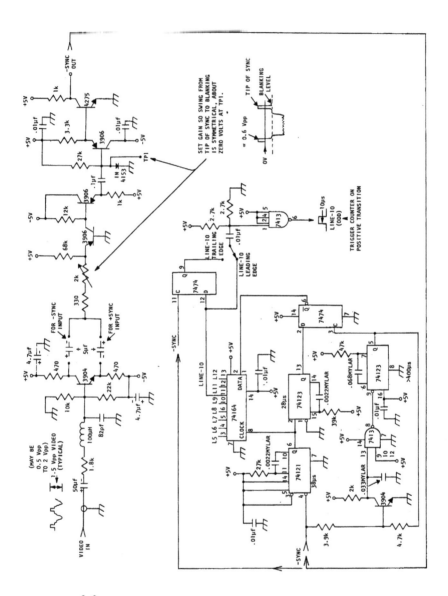

FIGURE 8.8.   SCHEMATIC OF TV LINE-10 EQUIPMENT DEVELOPED BY NBS.

152

Historically, many different radio signals have been used for distributing time and frequency. Some of the most successful adapted for this purpose have been those of the Loran-C navigation system. In the 1950's, after its initial development for navigation, work done at NBS showed that Loran-C signals could be quite stable, and since then, they have been controlled and used for frequency and time calibrations.

The success of Loran-C frequency and time signals is due to a number of things. Each station is controlled by cesium standards. Careful control and monitoring of the signals by the U. S. Coast Guard and the U. S. Naval Observatory allows a user to achieve rapid, accurate results. Groundwave signals provide state-of-the-art frequency and time calibrations. Even the skywave signals, though slightly less accurate, can become the basis of excellent high-quality calibrations. Loran-C coverage is increasing with the planned addition of new chains.

9.1   BASIC PRINCIPLES OF THE LORAN-C NAVIGATION SYSTEM

Loran-C is a radio navigation system that broadcasts on a frequency of 100 kHz using a bandwidth from 90 to 110 kHz. At this low frequency, the radio waves follow the earth's curvature; thus, they are relatively unbothered by the ionosphere and are very stable.

The Loran-C navigation system consists of many synchronized "chains" or networks of stations. These chains provide groundwave coverage of most of the United States, Canada, Europe, the North Atlantic, the islands of the Central and West Pacific, the Philippines, and Japan. Information about the various chains is given in Table 9.1.

One station in each chain is designated as a "master" station; the remaining stations are "slaves." The master station transmits groups of pulses that are received by the slave stations. The slave stations receive the master pulse groups and, at later times, transmit similar groups of synchronized pulses (figure 9.1).

On a ship or plane, the constant time differences between the reception of the master pulses and the corresponding slave pulses establish a line-of-position that is used for navigation. Signals from three separate Loran transmissions are needed to determine a line-of-position. For frequency and time applications, only a single Loran station is needed.

The coverage and accuracy of Loran-C signals depend upon the characteristics and format of the transmitted signal. When the system was designed, tradeoffs were made to make the system accurate and reliable. The characteristics and format of the transmitted Loran-C signals can be broken down into the following major elements:

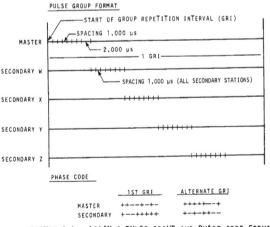

FIGURE 9.1.   LORAN-C PULSE GROUP AND PHASE-CODE FORMAT.

STATION LOCATION	STATION IDENTIFI- CATION		CODING DELAY (μs)	EMISSION DELAY (μs)
GREAT LAKES **REPETITION PERIOD:** **99,300 μs (SS7)**				
DANA, IN				
39° 51' .8" N 87° 29' 12" W	9930M (9960Z) *			
MALONE, FL				
30° 59' 39" N 85° 10' 9" W	9930W (7980M) *			14355.1
SENECA, NY				
42° 42' 51" N 76° 49' 34" W	9930X (9960M) *			31162.1
BAUDETTE, MN				
48° 36' 50" N 94° 33' 18" W	9930Y			
U. S. WEST COAST **REPETITION PERIOD:** **99,400 μs (SS6)**				
FALLON, NV				
39° 33' 6" N 118° 49' 56" W	9940M			

*These stations are members of two or more chains and are therefore dual or triple-rated. They transmit on two (or three) rates with two (or three) pulse repetition periods.

STATION LOCATION	STATION IDENTIFI-CATION	PEAK RADI-ATED POWER (kW)	BASELINE		PATH LENGTH (µs)	CODING DELAY (µs)	EMISSION DELAY (µs)
			NAUTICAL MILES	KILO-METERS			
MIDDLETOWN, CA 38° 46' 57" W 122° 29' 45" W	9940X	400		.	1,094.5	27,000	28094.5
SEARCHLIGHT, NV 35° 19' 18" N 114° 48' 17" W	9940Y	500			1,967.2	40,000	41967.3
NORTHEAST U. S. REPETITION PERIOD: 99,600 µs (SS4) SENECA, NY 42° 42' 51" N 76° 49' 34" W	9960M (9930X) *	800					
CARIBOU, ME 46° 48' 27" N 67° 55' 38" W	9960W (5930M)	350			2,797.2	11,000	13797.2
NANTUCKET, MA 41° 15' 12" N 69° 58' 39" W	9960X (5930X) *	275			1,969.9	25,000	26969.9
CAROLINA BEACH, NC 34° 3' 46" N 77° 54' 47" W	9960Y (7980Z) *	700			3,221.7	39,000	42221.7
DANA, IN** 39° 51' 8" N 87° 29' 13" W	9960Z (9930M) *	400			3,162.1	54,000	57162.1
SOUTHEAST U. S. REPETITION PERIOD: 79,800 µs (SL2) MALONE, FL 30° 59' 39" N 85° 10' 9" W	7980M (9930W) *	800					

*These stations are members of two chains and are therefore dual-rated. They transmit on two rates with two pulse repetition periods.

	STATION IDENTIFI-	EMISSION DELAY
STATION LOCATION	CATION	(μs)

WEST COAST OF CANADA
REPETITION PERIOD:
 59,900 μs (SH1)
─────────────

WILLIAMS LAKE, BC

   51° 57' 59" N   5990M
122° 22' 2" W

CENTRAL PACIFIC
REPETITION PERIOD:
 49,900 μs (S1)
─────────────

JOHNSTON ISLAND

  16° 44' 44" N   4990M
169° 30' 31" W

*These stations are members of two chains and are therefore dual-rated. They transmit on two
 rates with two pulse repetition periods.

156

STATION LOCATION	STATION IDENTIFI-CATION	PEAK RADI-ATED POWER (kW)	BASELINE		PATH LENGTH (µs)	CODING DELAY (µs)	EMISSION DELAY (µs)
			NAUTICAL MILES	KILO-METERS			
KURE, MIDWAY ISLAND 28° 23' 42" N 178° 17' 30" W	4990Y	300	849.6	1,573.4	5,253.1	29,000	34253.1
NORTH PACIFIC REPETITION PERIOD: 99,900 µs (SS1) ST. PAUL, PRIBILOFF IS., AK 57° 9' 10" N 170° 14' 60" W	9990M	275					
ATTU, AK 52° 49' 45" N 173° 10' 52" W	9990X	275			3,875.3	11,000	14875.3
POINT CLARENCE, AK 65° 14' 40" N 166° 53' 14" W	9990Y	1000			3,069.1	29,000	32069.1
NARROW CAPE, KODIAK IS., AK 57° 26' 20" N 152° 22' 11" W	9990Z (7960X) *	400			3,590.5	43,000	46590.5
NORTHWEST PACIFIC REPETITION PERIOD: 99,700 µs (SS3) WO JIMA, BONIN IS. 24° 48' 4" N 141° 19' 29" E	9970M	1000					
MARCUS ISLAND 24° 17' 8" N 153° 58' 52" E	9970W	1800	692.9	1,283.2	4,283.9	11,000	15283.9
HOKKAIDO, JAPAN 42° 44' 37" N 143° 43' 09" E	9970X	400		2,002.3	6,685.1	30,000	36685.1

hese stations are members of two chains and are therefore dual-rated. They transmit on two
ates with two pulse repetition periods.

STATION LOCATION	STATION IDENTIFI- CATION	CODING DELAY (μs)
GULF OF ALASKA REPETITION PERIOD: 79,600 μs (SL4)		
TOK, AK 63° 19' 43" N 142° 48' 32" W	7960M	
NARROW CAPE, KODIAK IS., AK 57° 26' 20" N 152° 22' 11" W	7960X (9990Z) *	11,000
NORTH ATLANTIC REPETITION PERIOD: 79,300 μs (SL7)		
ANGISSOQ, GREENLAND 59° 59' 17" N 45° 10' 27" W	7930M	
SANDUR, ICELAND 64° 54' 27" N 23° 55' 22" W	7930W (7970Y) *	
EJDE, FAROE ISLAND 62° 17' 60" N 7° 4' 27" W	7930X (7970) *	
CAPE RACE, NEWFOUNDLAND 46° 46' 32" N 53° 10' 28" W	7930Z (9930X) *	

*These stations are members of two chains and are therefore dual-rated. They transmit on two rates with two pulse repetition periods.

STATION LOCATION	STATION IDENTIFI- CATION	PEAK RADI- ATED POWER (kW)	BASELINE		PATH LENGTH (μs)	CODING DELAY (μs)	EMISSION DELAY (μs)
			NAUTICAL MILES	KILO- METERS			
NORWEGIAN SEA EPETITION PERIOD: 79,700 μs (SL3)							
JDE, FAROE ISLAND 62° 17' 60" N 7° 4' 27" W	7970M (7930X) ★	400					
BOE, NORWAY 68° 38' 7" N 14° 27' 47" E	7970X	250	654.7	1,212.5	4,048.2	11,000	15048.2
YLT, F. R. GERMANY 54° 48' 30" N 8° 17' 36" E	7970W	275	657.6	1,217.8	4,065.7	26,000	30065.7
SANDUR, ICELAND 64° 54' 27" N 23° 55' 22" W	7970Y (7930W) ★	1500	476.2	882.0	2,944.5	46,000	48944.5
JAN MAYEN, NORWAY 70° 54' 53" N 8° 43' 59" E	7970Z	250	520.2	963.4	3,216.3	60,000	62216.3
EDITERRANEAN SEA EPETITION PERIOD: 79,900 μs (SL1)							
LLIA MARINA, ITALY 38° 52' 21" N 16° 43' 6" E	7990M	250					
LAMPEDUSA, ITALY 35° 31' 21" N 12° 31' 30" E	7990X	400	284.0	526.0	1,756.0	11,000	12756.0
RGA BARUN, TURKEY 40° 58' 21" N 27° 52' 2" E	7990Y	250	529.4	980.4	3,273.3	29,000	32273.3
ESTARTIT, SPAIN 42° 3' 36" N 3° 12' 16" E	7990Z	250	646.9	1,198.0	3,999.7	47,000	50999.7

ese stations are members of two chains and are therefore dual rated. They transmit on two tes with two pulse repetition periods.

1. Carrier Frequency
2. Pulse Shape
3. Pulse Group

4. Phase Code
5. Blink Codes
6. Spectrum

The characteristics and format of the Loran-C signals and the important tradeoffs which were made in their selection are discussed below.

9.2 BASIC LORAN-C FORMAT

The basic format of the Loran-C transmissions, for a single rate, which would be received at a point in the coverage area of a Loran-C chain, is shown in the line labeled "phase code" in figure 9.2. Signals from other chains (adjacent and distant) would be superimposed on these signals (because all Loran-C signals share a carrier frequency of 100 kHz) but would have different Group Repetition Intervals (GRI). Master transmissions are labeled "M" in the figure and consist of 8 pulses (represented by the short vertical

lines) separated by 1000 microseconds and a ninth pulse 2000 microseconds after the eighth. The secondary transmissions are labeled "S" in the figure and consist of 8 pulses separated by 1000 microseconds. These pulses have the general envelope shape shown in the lower right corner of figure 9.2, and are positively and negatively phase coded as shown in the middle of the figure. Phase coding, station occupancy bounds, GRI, standard track point, etc., will be discussed in greater detail.

NOTE: The following information is included here for those readers who have a need to understand Loran-C operation in detail. An average frequency and time user may not wish to pursue this subject in such depth and may skip to section 9.3 on page 166.

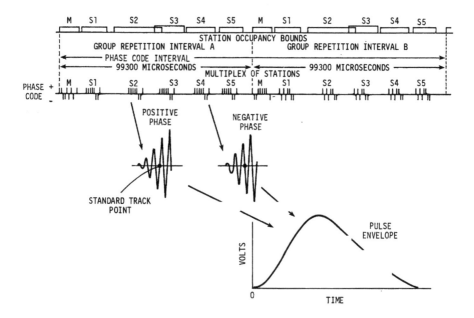

FIGURE 9.2. LORAN-C SIGNAL FORMAT (CHAIN OF SIX STATIONS AT GRI CODE 9930).

160

9.2.1 CARRIER FREQUENCY

All Loran-C transmissions are timeshared (time multiplexed) on a carrier frequency of 100 kHz. The choice of carrier frequency and the propagation medium determine the range to which the signals will be propagated (ground-waves and skywaves) and how stable they will be. High frequencies (greater than 200 kHz) have the best groundwave propagation characteristics. The best frequency for the Loran-C navigation system might have been in the band 150 to 200 kHz, but a frequency in this band was not available because it would have interfered with existing European broadcast stations. Loran-C signals at 100 kHz can be transmitted without inordinate difficulty, have good groundwave range and good stability. Loran-C transmissions may occupy a band from 90-110 kHz (see pulse shape).

9.2.2 TRANSMITTED PULSE SHAPE

The shape of the standard Loran-C pulse envelope as a function of time is given by:

$$f(t) = t^2 e^{-2t/\Delta t_p},$$

where Δt_p is the time to peak of the pulse and is 65 microseconds for a standard Loran-C pulse.

This pulse rises to approximately 50% of its peak amplitude in 25 microseconds. The rise time of the Loran-C pulse should be as fast as possible to allow maximum pulse amplitude at the receiver before the pulse has been contaminated by the arrival of skywaves. A squarewave would rise more quickly, but the 20 kHz bandwidth allowed for Loran-C pulses would be exceeded. The standard Loran-C pulse offers a good compromise between fast rise time and required bandwidth. Other pulse shapes having faster rise times that meet bandwidth requirements have been proposed, but not adopted. Transmitting equipment limitations place additional constraints on the type of pulse selected.

9.2.3 TRANSMITTED PULSE

The Loran-C transmitted pulse shape is defined in terms of the base current in the transmitting antenna. The antenna base current, $i(t)$, is given by the following:

$$i(t) = 0, t < \tau,$$

$$i(t) = A \left(\frac{t - \tau}{\Delta t_p} \right)^2$$

$$\exp \left[-2 \left(\frac{t - \tau}{\Delta t_p} \right) \right] \sin (\omega_0 t + \phi),$$

where:

τ $\leq t \leq t_p$

A is a constant related to the peak current (in amperes)

t is time (in microseconds)

τ is the time origin for the envelope; also called ECD (in microseconds)

t_p is the time of the pulse envelope peak ($65 + \tau$)

Δt_p is the rise time of the pulse envelope (65 microseconds)

ω_0 is the angular carrier frequency (0.2π radians/microsecond)

ϕ is the phase code (0 or π radians, $\pm 180°$)

For time (t) greater than t_p, $i(t)$ is controlled to satisfy radiated spectrum requirements. To prevent contamination of the leading edge of a Loran-C pulse by the tail of the previous pulse, ideally, the amplitude of the tail should be well attenuated before the next pulse starts. If the skywave of the tail of the pulse is considered, then a Loran-C pulse should be attenuated as fast as possible after attaining peak amplitude.

Unfortunately, the bandwidth constraint must be considered. A compromise between these two requirements is to allow a pulse length of 500 microseconds. By requiring the amplitude of the pulse at 500 microseconds to be down 60 dB from the peak of the pulse, the bandwidth limitation can be met and the pulse-tail skywave contamination problem can be avoided. Precise timing control of the leading edge of the Loran-C pulse is required to permit users to obtain accurate navigational information.

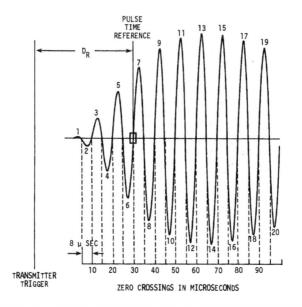

FIGURE 9.3.   STANDARD LORAN-C PULSE (ECD = 0, POSITIVELY PHASE-CODED).

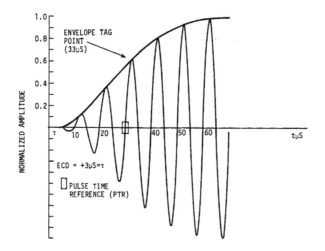

FIGURE 9.4.   LORAN-C PULSE WITH ECD = +3.0 MICROSECONDS.

9.2.4 GRAPHICAL REPRESENTATION

A graphical representation of the standard Loran-C pulse is shown in figure 9.3. The main aspects of the pulse are defined as follows:

Pulse Time Reference (PTR):

The positive-going zero crossing nearest to the 30 microsecond point on the pulse envelope for a positively phase coded pulse (negative-going zero crossing for a negatively phase coded pulse). In receivers, this point is often referred to as the "Track Point."

Reference Delay (D_R):

The fixed delay between transmitter triggers and the PTR.

Pulse Start:

The time 30 microseconds minus the algebraic value of the ECD earlier than the PTR.

Time-to-Peak (ΔT_p):

The time interval from pulse start to the peak value of the envelope of the pulse. ΔT_p is 65 microseconds.

Pulse Tail:

The portion of the pulse after the peak has occurred.

Zero Crossing:

The instant at which the direct coupled voltage (or current) reaches its zero DC level.

A. Envelope-to-Cycle Difference

Envelope-to-Cycle Difference (ECD) is the displacement between the start of the Loran-C pulse envelope and a zero crossing of the Loran-C carrier. ECD arises because the phase (carrier) velocity and the group (envelope) velocity of the Loran-C signal differ. As the signal propagates over sea water, the ECD decreases because the phase velocity exceeds the group velocity. ECD shifts over land paths are not always well behaved. The ECD of the transmitted signal is controlled by pre-distorting the transmitting antenna current pulse so that a signal with the desired ECD is obtained in the near far-field. ECD is illustrated in figure 9.4 (ECD = +3 microseconds). In the transformation between the transmitting antenna base current and the electric field in the far field, the carrier advances approximately 2.5 microseconds (because of the imaginary term in the theta component of the electric field vector) which results in a shift in ECD of +2.5 microseconds. In general, the ECD of the antenna base current waveform is adjusted to zero such that the ECD in the near far-field is +2.5. In general, this ECD decreases as the signal propagates such that it is approximately zero in the prime service area. ECD is an important parameter because if it is too great a

TABLE 9.2. ECD TABLE: HALF CYCLE PEAK AMPLITUDE vs ECD

HALF CYCLE NO.	NOMINAL TIME OF HALF CYCLE μs	ECD (μs)								
		-4	-3	-2	-1	0	+1	+2	+3	+4
1	+2.5	+.066	+.050	+.036	+.025	+.016	+.009	+.004	+.001	000
2	+7.5	-.166	-.144	-.122	-.102	-.083	-.066	-.050	-.036	-.025
3	+12.5	+.290	+.264	+.239	+.214	+.190	+.166	+.144	+.122	+.102
4	+17.5	-.419	-.393	-.367	-.341	-.315	-.290	-.264	-.239	-.214
5	+22.5	+.545	+.520	+.496	+.470	+.445	+.419	+.393	+.367	+.341
6	+27.5	-.659	-.637	-.615	-.592	-.569	-.545	-.520	-.496	-.470
7	+32.5	+.758	+.740	+.721	+.701	+.680	+.659	+.637	+.615	+.592
8	+37.5	-.840	-.825	-.810	-.793	-.776	-.758	-.740	-.721	-.701

*NORMALIZED TO PEAK AMPLITUDE (tp = 65 μs); i.e., AMPLITUDE AT t_p = 1.

163

receiver will not be able to identify the proper "Track Point."

B. ECD Determination

A standard measurement process has been defined to determine the ECD of the Loran-C pulse. Amplitude samples of the first eight half cycle peaks or quadrature samples at 2.5 microsecond intervals about the PTR through the first forty microseconds of the pulse are taken. A best-fit standard Loran-C pulse is determined for the measured pulse using a minimum mean square error (MMSE) estimator. A cost function which is a function of both peak pulse amplitude (A) and ECD (τ) is minimized. This cost function is minimized by solving two partial differential equations. The peak amplitude and ECD of the measured data are found from this best-fit pulse. An rms pulse shape error computation is associated with this determination of ECD. Confidence limits on ECD are being established for various magnitudes of rms pulse error. At Loran-C transmitting stations, the ECD is determined by measuring half cycle peak amplitudes through the first forty microseconds of the pulse.

C. Pulse Group

Loran C receivers use carrier phase measurements of the pulsed 100 kHz signals. High accuracy is achieved by allowing separation of the groundwave and skywave components of the Loran-C signal due to pulse modulation. Because all chains operate on the same frequency, some method of chain selection is necessary. This is achieved by transmitting signals from different chains on different group repetition intervals (GRI) (see figure 9.2). In selecting the number of pulses to transmit per GRI and their spacing, tradeoffs had to be made to maximize the number of pulses per second which could be received (to increase effective receiver signal-to-noise ratio (SNR) for a given peak power), to minimize interference from first-hop skywaves and multi-hop skywaves, and to provide sufficient time for receiver recovery (processing).

D. Existing Format

The stations in a particular Loran-C chain transmit groups of pulses at a specified group repetition interval (GRI). Possible

TABLE 9.3. GROUP REPETITION INTERVALS

(GRI In Tens of Microseconds)

9999		8999		7999		6999		5999		4999	
9998		8998		7998		6998		5998		4998	
9997		8997		7997		6997		5997		4997	
•		•		•		•		•		•	
•		•		•		•		•		•	
9991		8991		7991		6991		5991		4991	
9990	(SS1)	8990		7990	(SL1)	6990		5990	(SH1)	4990	(S1)
9989		8989		7989		6989		5989		4989	
•		•		•		•		•		•	
•		•		•		•		•		•	
9971		8971		7971		6971		5971		4971	
9970	(SS3)	8970		7970	(SL3)	6970		5970	(SH3)	4970	(S3)
9969		8969		7969		6969		5969		4969	
•		•		•		•		•		•	
•		•		•		•		•		•	
•		•		•		•		•		•	
9931		8931		7931		6931		5931		4931	
9930	(SS7)	8930		7930	(SL7)	6930		5930	(SH7)	4930	(S7)
9929		8929		7929		6929		5929		4929	
•		•		•		•		•		•	
9000		8000		7000		6000		5000		4000	

values for GRI range between 40,000 microseconds (25 groups per second) and 100,000 microseconds (10 groups per second) in steps of 100 microseconds. The master station transmits a total of nine pulses as shown in figure 9.2. The first eight pulses are separated by 1000 microseconds and the ninth pulse is separated from the eighth by 2000 microseconds. The ninth pulse is used for identification of the master signal. Secondary stations transmit eight pulses which are separated by 1000 microseconds.

E. GRI

The GRI in Loran-C chains are selected to improve receiver operation while reducing interference from adjacent chains (cross rate interference). The GRI must be long enough to prevent overlap of the master and any secondary signals in the coverage area. Time must be allowed for between the last secondary's eighth pulse and the master's first pulse in the next group.

Pulses Per Group

With multiple pulsing in a GRI, it is possible to increase the effective average pulse rate and effective average power without increasing the peak power or losing the advantages of pulse transmission and timesharing. Each doubling of the number of pulses radiated results in twice the previous average radiated power (3 dB) and increases the far-field strength by the square root of two. Thus, going from one to eight pulses results in 9 dB more average power. In this manner, the range of the system is maximum for a given peak power. Groups of pulses in the series 2, 8, 16, etc., provide the best phase codes for skywave rejection.

Pulse Spacing

Loran-C pulse spacing is chosen to be 1000 microseconds to provide enough time for first or second hop skywaves to decay before the next pulse arrives.

Phase Coding

Phase coding serves several purposes:

(1) It is a means to identify master and secondary stations automatically.

(2) It allows for the automatic search process.

(3) It rejects multi-hop skywaves.

Through the use of phase coded signals and cross-correlation techniques in receivers, automatic identification of the master's and secondaries' signals is possible. Phase coding extends over a period of two GRI (called the Phase Code Interval (PCI) as shown in figure 9.2. Master and secondary phase codes are shown in figure 9.5. Basically, when skywaves, whose phase codes differ from the desired groundwaves, are received during receipt of the groundwave, the skywave signal is cancelled.

	STATION	
GRI	MASTER	SECONDARY
A	+ + - - + - + - +	+ + + + + - - +
B	+ - - + + + + -	+ - + - + + - -

FIGURE 9.5. LORAN-C PHASE CODES.

F. Blink Codes

Blinking is used to warn navigation users that there is an error in the transmissions of a particular station. Blink is accomplished

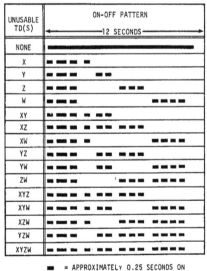

FIGURE 9.6. LORAN-C BLINK CODE.

165

at the master station by turning the ninth pulse on and off in a specified way as shown in figure 9.6. The secondary station of the unusable pair blinks by turning its first two pulses on and off. The first two pulses of the secondary station are turned on (blinked) for approximately 1/4 second every 4 seconds. All secondaries use the same code. Most modern navigation receivers automatically detect secondary station blink only as this is sufficient to trigger the alarm indicators.

G. Spectrum

The transmitted spectrum is defined to be the spectral density of the time function f(t), where f(t) is

$$t^2 e^{-2t/65}.$$

The spectral density function is given by

$$S(\omega) = \left[\frac{|F(\omega)|}{|F(\omega)|_{max.}} \right]^2$$

where:

$$F(\omega) = \int_{-\infty}^{\infty} f(t) \, \varepsilon^{-j\omega t} \, dt$$

and

$$\omega = \text{angular frequency (rad/sec)}.$$

The Loran-C spectrum follows provisions of the Radio Regulations of the International Telecommunications Union (ITU). The total energy outside the 90 - 110 kHz band is less than 1% of the total radiated energy. The energy below 90 kHz and above 110 kHz are no greater than 0.5% of the total radiated energy. As a practical matter for Loran-C type signals, this is achieved if the spectral density of the radiated signal at 90 kHz and 110 kHz is down at least 20 dB relative to its value at 100 kHz.

H. Fine Spectra

There are several hundred spectral lines in the near band spectrum of the Loran-C signals. These spectral lines result from the various periodic elements of the Loran-C transmissions. Among the most significant are lines every:

1 kHz - 1000 microsecond pulse spacing

125 Hz - Blocks of 8 pulses with 1000 microsecond spacing

1/GRI Hz - Group repetition interval

1/(2 x GRI) Hz - Phase code interval

NOTE: When observing the spectrum on analyzers whose minimum IF bandwidths are greater than the spacing between the spectral lines to be observed, the fine spectra will be a result of the spectrum analyzer's IF response rather than the Loran-C signal.

I. Harmonics

To assure that Loran-C transmissions do not interfere with other services in adjacent parts of the radio spectrum, the spectral level of any harmonic relative to the value at 100 kHz doesn't exceed the following levels:

HARMONIC	LEVEL
2nd	-70 dB
3rd	-80 dB
4th	-85 dB
5th (or greater)	-90 dB

9.3 WHAT IS THE EXTENT OF LORAN-C COVERAGE?

There are two methods of signal propagation: groundwave and skywave. Using the groundwave, Loran-C timing is possible to about 1500 miles over land and 2000 miles over the ocean. Skywave reception beyond 5000 miles is possible.

9.3.1 GROUNDWAVE SIGNAL RANGE

Radio energy from each Loran-C transmitter radiates in all directions. A portion of the energy travels out from each transmitting station parallel to the surface of the earth. This is the groundwave.

Useful Loran-C groundwave coverage extends from approximately 1500 to 2000 miles. During periods of good reception, this range may be greater and during periods of high noise and interference, it may be less.

However, based on typical noise and interference situations, 1500 miles is a reasonable estimate of the <u>reliable</u> groundwave range of Loran-C signals from a station having 300 kilowatts peak pulse power.

The ability of the Loran-C timing receiver to indicate accurate time readings is dependent upon the relative strength of the received signals and the level of the local noise interference. As the signal-to-noise ratio decreases, the performance of the receiver decreases. As a result, accuracy deteriorates at receiver locations far away from the stations. For timing purposes, long-term averaging may be employed to reduce errors.

9.3.2 SKYWAVE SIGNAL RANGE

Some portion of the Loran radio signal radiates upward from the transmitting antenna and is reflected from the electrified layer of the atmosphere known as the ionosphere. This signal is called the skywave. If the groundwave signal has traveled a long distance, it will be reduced in amplitude and weakened. The receiver will pick up both the groundwave and the skywave signals. The receiver cannot easily lock to the groundwave because it is weak and noisy, so it will lock to the skywave signal.

But the skywave signal "moves" around because it is reflected off the ionosphere. This movement is caused by the corresponding motion of the ionosphere due to the action of the sun. The most pronounced effects occur at sunrise and sunset, but there are also smaller changes throughout the day and night. Each of these path changes looks like a change in the signal phase at the receiving point.

Because the skywave signals are stronger than groundwave signals at great distances, skywave lock-on is possible at distances beyond where the groundwave signals can be received. A strong skywave signal allows the receiver to average and reproduce its skywave phase plot day after day. What we are saying is that, even though the skywave signal moves around, it does so in a predictable way. With good record-keeping, the skywaves can yield very good time and frequency data.

The next section will explain how Loran works and how it can be used for timing. Keep in mind that Loran can supply high accuracy frequency and time calibrations, but under weak or noisy signal conditions, cycle selection is difficult and requires a skilled operator. If this extreme timing accuracy is not required, other techniques should be examined.

9.4 WHAT DO WE GET FROM LORAN-C?

Loran-C uses several modulation techniques to improve its use for navigation. These translate directly into an ability to disseminate both frequency and time with greater accuracy. Foremost among its characteristics is the fact that the signals are transmitted as pulses--and pulses at a carrier frequency as low as 100 kHz are unusual.

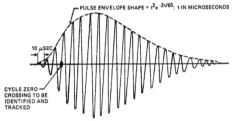

FIGURE 9.7. THE 100 kHz LORAN-C PULSE.

Why does Loran use pulses? The accuracy of the system is based on the fact that the early part of the pulse--the one that leaves the transmitter first--will travel along the ground and arrive before the less stable skywave pulse that bounces off the ionosphere-- so, the groundwaves and skywaves can be sorted out at the receiver. If your location or reception conditions do not permit groundwave reception and use, the timing accuracy achieved by using skywaves may be many times worse. But, even with skywaves, frequency calibration accuracy is still very good.

9.4.1 SIGNAL CHARACTERISTICS

In addition to transmitting pulses for separation of ground and skywave signals, the pulses are transmitted in groups for identification of the particular station being received. The slave stations in a particular Loran-C chain transmit eight pulses to a group. For identification, the master station transmits a ninth pulse in each group.

Loran station identification is also aided by separately phase-coding the master and slave pulses. Each group of pulses is coded by a phase reversal process which enables the receiver to eliminate certain types of interference. Also, since all Loran-C's broadcast at 100 kHz, it is possible to receive many different Loran chains at the same time. Therefore, the pulses in each chain are transmitted at slightly different rates so the receiver can distinguish between them.

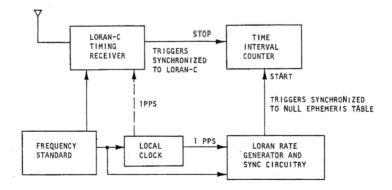

FIGURE 9.8. INSTRUMENTATION FOR UTILIZING PULSES WITHIN THE
LORAN-C PULSE GROUPS.

All these subtle differences in transmission pulse rates give rise to the need for a fairly complex receiving system. Although relatively simple Loran-C navigation equipment exists, the use of Loran-C for timing requires greater interaction by the operator. As might be expected, the cost of receiving equipment is higher than less complex VLF or high frequency radio equipment. Manual signal lock-on, which is required with some Loran-C receivers, means that the operator skill required is fairly high. Still, once the equipment is installed and the operator gains a little experience, the results can be exceptionally accurate. As a frequency and time calibration system, Loran-C is almost unequaled.

Figure 9.10 shows how signals from a Loran-C chain are timed. Plus and minus signs over individual pulses indicate the phase code, where (+) represents a relative phase of 0° and (-) represents 180°. All master stations are phase-coded one way, and all slaves are phase-coded another.

9.4.2 TIME SETTING

To make time calibrations, the user of Loran-C must take into account the fact that the pulse repetition rates used are not at one pulse per second, nor are they a convenient

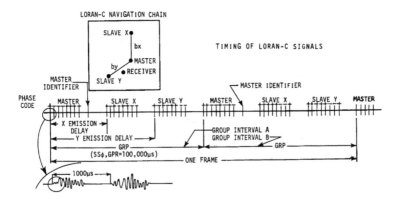

FIGURE 9.9. TIMING OF LORAN-C SIGNALS.

(A)

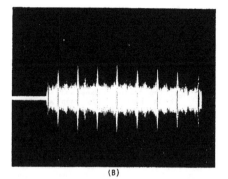

(B)

(C)

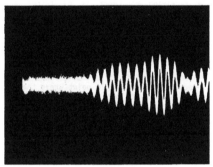

(D)

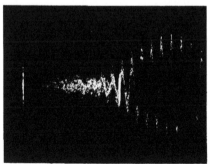

(E)

FIGURE 9.10. STEPS IN LOCKING ONTO A LORAN-C
SIGNAL: THE PULSES SHOULD BE
CLEARLY VISIBLE ABOVE THE NOISE,
AS IN (A). (B) MOVE THE PULSES
TO THE LEFT AND INCREASE 'SCOPE
SWEEP SPEED TO 1 ms/cm. (C) SHOWS
THE FIRST OF THE EIGHT PULSES AT A
FASTER 'SCOPE SWEEP SPEED. NOTE
THAT THE SINE WAVES ARE DOUBLED,
INDICATING LACK OF CODING. (D)
RESOLVE THIS BY ADJUSTING RECEIVER.
(E) SHOWS THE FINAL STAGE OF PULSE
POSITIONING WITH THE BRIGHT SPOTS
ON THE WAVEFORM AS THE RECEIVER
TRACKING POINT.

169

multiple thereof. Instead, a time of coincidence, TOC (when a Loran pulse coincides with a one-second pulse) may occur as infrequently as every sixteen minutes. After the receiver is locked to the incoming signal, the one pulse-per-second timing output must be synchronized at the time of coincidence. This is done by pushing a button on the receiver just before the correct second arrives. Assuming power is not lost, and the receiver does not lose lock, the timing pulses will stay "on time." It is important to keep a battery supply connected to the receiver to prevent loss of synchronization in case of power outages.

Loran-C is most often used when the required timing accuracies are equal to or better than ten microseconds. Remember that any user of a timing system that provides accuracies of this order must take into account path and equipment delays. In addition, he must pay very close attention to how the receiving equipment is maintained and have the necessary battery backups. Simply stated, as with any scheme for time and frequency calibrations, if you are looking for high accuracies, you must be willing to pay the price in equipment complexity and operator attention.

The most difficult part of Loran-C timing is recognizing the sample point where the receiver is locked onto the signal. Since Loran-C operates at 100 kHz, it is possible to lock onto carrier zero crossings separated by the 10 microsecond period. An oscilloscope display of Loran-C pulses is helpful and under noisy conditions, it is of great assistance to the operator to have a graphical plot made of the received pulse showing the exact position of the sample point. Some receiving equipment sold for Loran timing usually incorporates this plotting capability.

Setting Local Time.

In a timing system, it is often convenient or necessary to operate "on time" rather than simply with a known time offset. The equipment needed is basically a clock that has a display for the time-of-day readout.

The following steps show how to synchronize local time by Loran-C:

1. Obtain a coarse clock synchronization to UTC by reference to WWV. Allow for propagation and receiver delays if known.

2. Acquire and track a Loran-C transmission from either a slave or master transmitter.

3. On an upcoming time-of-coincidence (TOC), operate the clock-sync controls so that clock circuits will synchronize the internal receiver time signal when the TOC occurs.

4. Using Loran-C emission and propagation delay figures (available, along with TOC tables, from USNO, Time Services Division, Washington, DC 20390) and the receiver delay from the operating manual, the clock time can be corrected to be "on time."

This completes synchronization. Subsequent "one-second" pulses from the receiver will be "on time."

9.4.3 FREQUENCY CALIBRATIONS USING LORAN-C

Frequency calibration using Loran-C is less complicated than time recovery, but the operator must still acquire the station and lock the receiver to a correct operating point. Under noisy reception conditions, the oscilloscope display of the received pulses may be too noisy to allow the tracking point to be positioned accurately. Unless the operator can identify eight pulses, the receiver may fail to lock because of the inherent phase coding.

Assuming that lock-on has been acquired, the receiver will track the local oscillator against the received signal and produce a phase plot. The chart then simply represents the beat note between a local standard and the received Loran signals.

How to Measure Frequency.

The frequency of a local frequency standard is measured in two steps. First, a phase-tracking record is prepared using the signal radiated by a synchronized Loran-C transmitter. The record should cover a period that is appropriate to the desired measurement accuracy. For example, in most locations, a groundwave phase record extending over twenty-four hours is adequate for relative frequency determinations having a probable error of one or two parts in 10^{11} to 10^{12}. Shorter periods may be used--or the skywave rather than the groundwave may be tracked--when extreme accuracy is not needed.

The second step is to reduce the data provided by the record to yield frequency difference. The basic expression used for this purpose, as discussed in Chapter 3, is:

$$\frac{\Delta f}{f} = -\frac{t_2 - t_1}{T},$$

170

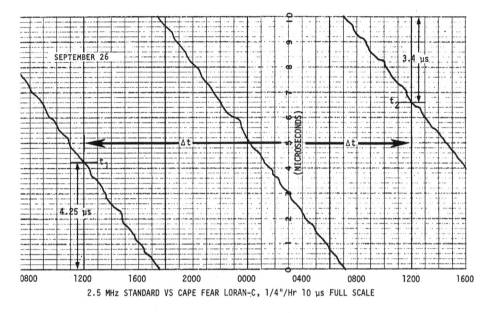

2.5 MHz STANDARD VS CAPE FEAR LORAN-C, 1/4"/Hr 10 μs FULL SCALE

FIGURE 9.11.   SAMPLE RECORD OF RECEIVED LORAN-C SIGNAL, NOON SEPTEMBER 26
THROUGH NOON SEPTEMBER 27, 17.65 MICROSECONDS.

where $\Delta f/f$ is the relative frequency of the local frequency source, and t_1 and t_2 are initial and final phase or time differences, respectively, obtained from the record over the averaging time interval, T.

· A sample phase record is shown in figure 9.12. Here we see that the relative phase between the local standard and the Cape Fear Loran-C signal stood at t_1 = 4.25 microseconds at noon. At the following noon (86,400 seconds later), the relative time had moved across the chart once and then to 5.9 microseconds. Thus, the relative frequency difference for the indicated twenty-four hour period was:

$$\frac{\Delta f}{f} = -\frac{\Delta t}{T} = -\left(\frac{-(3.45 + 10 + 4.1)}{8.64 \times 10^{10}} \right)$$

$$= 2.03 \times 10^{-10}$$

9.5   HOW GOOD IS LORAN-C?

The frequency and time services provided by Loran-C are capable of providing calibra-tions at very high accuracy levels. Of course the accuracy achieved will depend on propagation conditions, the path from the transmitter to the user, and the length of time over which the measurements are averaged.

At this writing, Loran-C is one of the principal methods of intercomparing the atomic standards at the NBS, the USNO and the National Research Council in Canada. This scheme is also extended to the European Loran network and affords a means for the BIH (Paris) to compare all of the world's main time scales. The results reported for this system of intercomparison are impressive--approaching 1 part in 10^{13} over several months.

Typical values for groundwave reception over land could easily approach a few parts in 10^{12} for a one-day average. Corrections to Loran data are provided by the USNO, and the NBS also monitors and reports daily Loran phase readings. Users who want very accurate results from Loran-C must correct the observed data by using the published data. This information is available by TWX on a daily basis from the USNO or by mail weekly from the USNO and monthly from NBS.

9.5.1 GROUNDWAVE ACCURACY

For timing, the accuracy of the ground-wave at a particular location depends upon several factors that contribute to the errors in the measurements. The approximate errors are:

1. Error in transmitted signal, 0.2 microseconds.

2. Errors caused by propagation path, up to 0.2 microseconds over water; up to 1 microsecond over land.

3. Receiver errors, 0.02 microsecond typical.

4. Errors caused by atmospheric noise and interference, up to 0.5 micro-second for interference. Longer averaging times can be used with noisy signals.

9.5.2 SKYWAVE ACCURACY

At frequencies near 100 kHz, skywave stability depends on the reflecting medium-- the ionosphere. Frequently, the ionosphere causes small signal amplitude fades or phase variations for long periods during total day-light on the path. As a general rule, expected one- or two-hop skywave stability is approximately several microseconds and is quite usable for timing.

9.6 ARE THE DATA VALID?

The Loran-C groundwave, unlike other VLF or LF transmissions that are used for frequency calibration, has no characteristic phase signature such as the diurnal phase shift of the skywave at VLF or the WWVB-type of periodic phase offset and return. If, in the absence of "built-in" phase disturbances of this type, substantiation of the Loran phase record is desired, it can be accomplished in two ways:

1. Some Loran receivers provide for a signal envelope record showing the position of tracking points that can be made on a second chart recorder. See figure 9.13.

2. A concurrent record of signal ampli-tude can be made. However, this does not provide evidence of track-ing as clearly as the recycling envelope trace suggested above.

Shown are two plots of the received Loran pulse. Notice the sample point where the

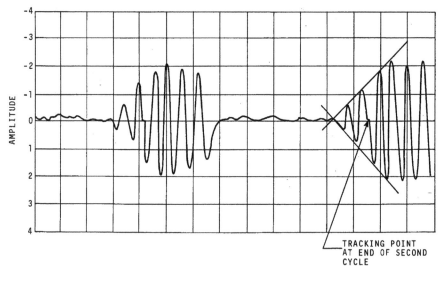

FIGURE 9.12.   PLOT OF THE RECEIVED LORAN-C SIGNAL.

172

receiver is tracking the signal. Although this method does not always guarantee cycle count, it at least will allow the operator to reset the receiver at the same point after the signal is lost and reacquired.

9.7 SUMMARY

Loran-C broadcasts provide an excellent source of signals for both frequency calibrations and time recovery. Loran may not be the best choice for all users, especially those who have only routine low-to-medium accuracy calibrations to perform. Balanced against this, however, is the fact that there are to be a number of chains in operation and the signal strength will be high. It can also be expected that more Loran manufacturers will enter the marketplace. This should provide lower cost instruments that are easier to operate.

Like all other services described in this book, Loran-C requires that the operator expend more effort for very accurate results. This is especially true for accurate time recovery--the user must not confuse the resolving power of Loran-C (in the nanosecond region) with the accuracy (1 to 2 microseconds) that the average user can achieve. It is also worth repeating the suggestion to any potential user of Loran-C (or any other service) to use the manufacturer's published material. If any doubts exist, try to obtain a receiver on loan and gain some first-hand experience before investing in any technique for time and frequency calibration.

For many years people have realized that an earth satellite would be the nearly ideal way to broadcast radio signals for time and frequency calibrations. At this writing, two systems are operational and another is being planned.

NBS provides a continuous time code via the GOES (Geostationary Operational Environmental Satellite) satellites. The U. S. Navy operates the TRANSIT series of navigation satellites with major participation by the Applied Physics Laboratory of the Johns Hopkins University and the U. S. Naval Observatory. The operation of these two systems will be explained in this chapter.

In addition to GOES and TRANSIT, the proposed Global Positioning Satellite System (GPS) is in the experimental stage. When it becomes operational, it will offer users several advantages. First, its frequency band, signal strength, and the use of state-of-the-art atomic oscillators on board each of the planned 24 satellites will allow high-accuracy calibrations. Second, its predicted popularity should make available a line of equipment that will be versatile and easy to use. However, since no firm operational date can be specified, GPS will not be covered in this chapter.

10.1   THE GOES SATELLITE TIME CODE

10.1.1   BACKGROUND

A satellite signal is no different than any other radio signal so far as the user is concerned. Table 10.1 gives the main specifications for GOES. Note that it broadcasts in the UHF band (which means small antennas are used), the system supports two satellites at this time, and the satellites are nearly stationary. That is, if you point the antenna to one part of the sky, you don't have to move it to keep a satellite in view.

The user should not be discouraged by the difference in frequency or the operating techniques needed for frequency and time calibrations using GOES. In fact, tests have shown that GOES is one of the easiest services to use and that it can be used in areas where other signals are nearly impossible to receive. The equipment provided by manufacturers is usually quite sophisticated and nearly automatic in its operation.

GOES is an operational descendant of NASA's Applications Technology Satellites.

TABLE 10.1   SPECIFICATIONS FOR THE GOES SATELLITES

FREQUENCY:	468.8250 MHz (Western GOES)
	468.8375 MHz (Eastern GOES)
POLARIZATION:	RIGHT-HAND CIRCULAR
MODULATION:	CPSK (± 60°)
DATA RATE:	100 BPS
SATELLITE LOCATION:*	135° W (Western GOES)
	75° W (Eastern GOES)
SIGNAL STRENGTH AT EARTH'S SURFACE (OUTPUT FROM ISOTROPIC ANTENNA):	-139 dBm
CODING:	MANCHESTER
BANDWIDTH:	400 Hz

*Other satellite locations may be used occasionally when operations must be switched to in-orbit spare satellites.

The GOES satellites are in orbit 36,000 kilometers above the equator. They travel at about 11,000 kilometers per hour and remain continuously above nearly the same spot on earth. They are thus termed geostationary. Since they have the same regions of earth in constant view, they can provide 24-hour, continuous service.

The advantages of geostationary satellites for time broadcasting are numerous. They are almost always in view and provide a source of continuous synchronization. Non-geostationary satellites, on the other hand, offer exposure to the user only for short periods of time at intervals ranging from about one hour to many hours or even days. Thus, non-geostationary satellite systems usually have a number of satellites in orbit.

Because of the bandwidth and its center frequency, the GOES system operates on the technique of using time measurements to calibrate frequency. As other chapters in this book explain, a time variation can be translated into a frequency difference. When the user buys a GOES receiver, he will have an instrument that expects the user to connect his 1 pps clock for comparison. This implies that the user must have such a clock and maintain it with batteries, etc., in order to use the GOES signals.

There is nothing to prevent manufacturers from using frequency measurements directly and from building equipment to do this. However, GOES has a time-code output (perhaps its major application is for its time code) and since the manufacturers felt the largest number of users wanted the time code, they developed their equipment for time recovery. Frequency calibrations obtained using GOES time data fall into the accuracy range from 1 part per million to 1 part in 10^9 (for a 1-day measurement period) with respect to NBS, depending on the sophistication of the equipment used.

The path from the NBS master clock to the user via a geostationary satellite is, in effect, a line-of-sight path allowing the use of high carrier frequencies that are largely unaffected by the ionosphere and troposphere. This means negligible fading and path length variations which are so characteristic of terrestrial HF signals. The line-of-sight path also means that free space transit time computations will work for most timing applications.

The NBS master clock is located on the ground rather than in the satellite itself. The satellite is only a "bent pipe" or "transponder" used to relay signals. This allows for easy control and maintenance of the system, thus guaranteeing better performance and reliability.

Because of these advantages, the GOES satellite can provide a time message, repeated continuously, to clocks in its view. It can control the frequency rate of the slaved ground clock to eliminate the need for high-quality oscillators and can also provide position data to correct for propagation delays.

10.1.2 COVERAGE

There are three GOES satellites in orbit, two in operational status with a third serving as an in-orbit spare. The two operational satellites are located at 135° and 75° West Longitude and the spare is at 105° West Longitude. The western GOES operates on 468.825 MHz; the eastern on 468.8375 MHz. The earth coverages of the two operational satellites

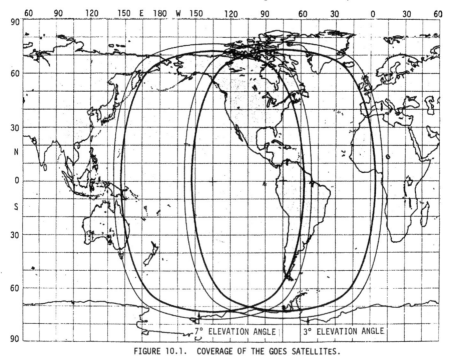

FIGURE 10.1. COVERAGE OF THE GOES SATELLITES.

176

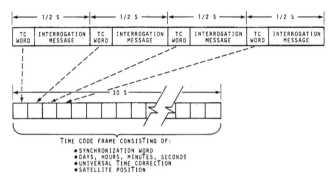

FIGURE 10.2 INTERROGATION CHANNEL FORMAT.

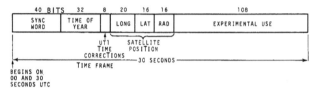

FIGURE 10.3 TIME CODE FORMAT.

are shown in figure 10.1. The heavy oval lines overlap the points on earth where each satellite is 7° above the horizon. The lighter lines represent a 3° elevation angle.

10.1.3 SIGNAL CHARACTERISTICS

The signal characteristics are summarized below. The two satellite signals are both right-hand circularly polarized and separated in frequency by 12.5 kHz. The data rate is 100 b/s requiring 400 Hz of bandwidth. The data are Manchester coded and phase modulate the carrier ± 60°, thus providing a carrier for the application of conventional phase-lock demodulation techniques.

The main purpose of the GOES satellites is to collect environmental data from remote ground sensors. The time code was added to provide time and a date to the collected data.

10.1.4 TIME CODE FORMAT

The time code is part of the interrogation channel which is used to communicate with the remote environmental sensors. Interroga-

tion messages are continuously being relayed through the GOES satellites. The format of the messages is shown in figure 10.2.

Each interrogation message is one-half second in length or 50 bits. The data rate is 100 b/s. The time-code frame begins on the one-half minute and takes 30 seconds to complete (see figure 10.3). Sixty interrogation messages are required to send the 60 BCD time-code words constituting a time-code frame.

The time-code frame contains a synchronization word, a time-of-year word (UTC), the UT1 correction, and the satellite's position in terms of its latitude, longitude, and height above the earth's surface minus a bias of 119,3000 microseconds. The position information is presently updated every 4 minutes.

As shown in figure 10.2, an interrogation message contains more than timing information. A complete message consists of four bits representing a BCD time code word followed by a maximum length sequence (MLS) 15 bits in length for message synchronization, and ends with 31 bits as an address for a particular remote weather data sensor.

177

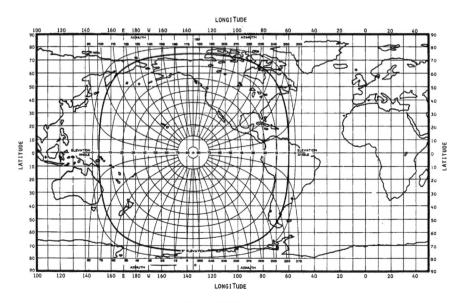

FIGURE 10.4.   WESTERN SATELLITE POINTING ANGLES.

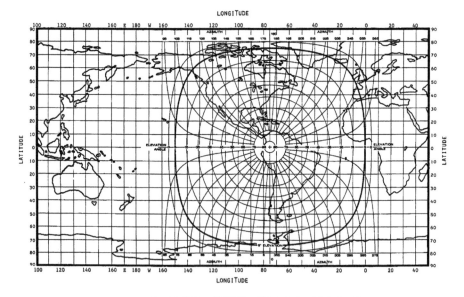

FIGURE 10.5.   EASTERN SATELLITE POINTING ANGLES.

178

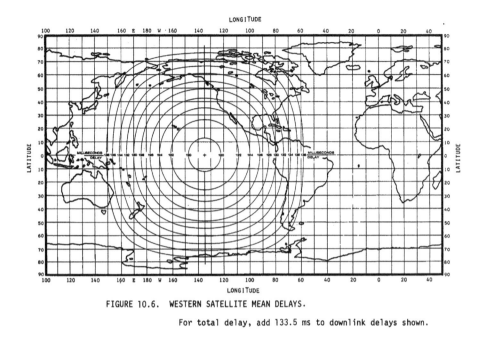

FIGURE 10.6.   WESTERN SATELLITE MEAN DELAYS.

For total delay, add 133.5 ms to downlink delays shown.

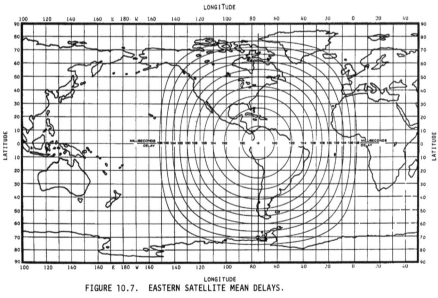

FIGURE 10.7.   EASTERN SATELLITE MEAN DELAYS.

For total delay, add 124.5 ms to downlink delays shown.

10.1.5 ANTENNA POINTING

Pointing an antenna to either satellite is relatively simple. Because of the large beamwidths of low-gain antennas (< 10 dB), pointing into the general direction of the satellite is usually sufficient. However, the antenna must be located so it has an unobstructed path to the satellite. Figures 10.4 and 10.5 provide detailed information for antenna pointing with elevation and azimuth angles provided. For example, the pointing directions to the eastern satellite from San Francisco can be obtained from figure 10.5 as approximately 119° azimuth and 24° elevation. Satellite positions may change with time. Users should check the monthly NBS Time and Frequency Bulletin for current status.

10.1.6 PERFORMANCE

The GOES time code can be used at three levels of performance: uncorrected, corrected for mean path delay only, and fully corrected.

A. Uncorrected

The path delay from point of origin (Wallops Island, Virginia) to the earth via the satellite is approximately 260,000 microseconds. Since the signals are advanced in time by this value before transmission from Wallops Island, they arrive at the earth's surface nearly on time (within 16 milliseconds).

B. Corrected for Mean Path Delay

Accounting for the mean path delay to any point on the earth's surface, but ignoring the cyclic (24-hour) delay variation, generally guarantees the signal arrival time to ± 0.5 millisecond. For example, the mean delay to San Francisco through the eastern satellite is 130.5 milliseconds downlink (from figure 10.7) plus 124.5 milliseconds uplink delay, for a total of 255 milliseconds. Since the time is advanced by 260 milliseconds before leaving Wallops Island, the time arrives at San Francisco 5 milliseconds early.

C. Fully Corrected

The cyclic path delay variation on GOES is a result of the satellite orbit or path around the earth not being perfectly circular and not in the plane of the equator. The orbit is actually an ellipse and has a small inclination--usually less than 1°. To compen-

sate for these and other effects, the satellite position is included with the time message for correction of path delay by the user. (See [18] and [19] for details on how this is accomplished.) This correction provides path delays accurate to ± 10 microseconds under normal operating conditions. The ultimate accuracy of the recovered time depends upon knowledge of user equipment delays and noise levels as well as path delay.

Typical delay variations for the eastern satellite are shown in figure 10.8. The peak-to-peak values change as the satellite's inclination and eccentricity are varied by orbit maneuvers and natural perturbations.

10.1.7 EQUIPMENT NEEDED

A. Antennas

Many antennas have been used by NBS with varying results. The best performance has been obtained using a right-hand circularly polarized helix antenna (fig. 10.9) with about 10 dB gain. Excellent results have been obtained with the microstrip antenna shown in figure 10.10. Dipoles and loops have also worked at lower levels of performance. All of these antennas had gains in the range of 3 to 10 dB.

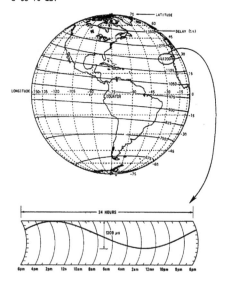

FIGURE 10.8 TYPICAL DELAY VARIATIONS FOR THE EASTERN SATELLITE.

180

FIGURE 10.9.   RIGHT-HAND CIRCULARLY POLARIZED
HELIX ANTENNA USED TO RECEIVE
GOES SIGNALS.

FIGURE 10.10.   MICROSTRIP ANTENNA USED TO
RECEIVE GOES TIME CODE.

B. Receivers

The NBS-built receiver shown in figure 10.11 is completely automatic, requiring no tuning or auxiliary equipment to operate. Commercial versions of this equipment are also available. For information on manufacturers, contact the Time and Frequency Services Group, 524.06, NBS, Boulder, CO 80303.

10.1.8 SATELLITE-CONTROLLED CLOCK

The clock shown in figure 10.11 was designed to interface with a receiver and achieve bit and frame synchronization, strip the time code from the interrogation message,

FIGURE 10.11.   GOES TIME CODE RECEIVER.

obtain time code frame sync, and read the time code message into memory. The clock then updates the memory time by counting cycles of the interrogation channel data clock, a 100 Hz square wave, and verifying its memory time through continual comparisons with consecutive time code messages from the satellite [20].

This clock was designed to address the majority of user's needs at minimum cost. This clock was not designed to "squeeze" the last bit of performance from the GOES timing system. The clock does not directly use the satellite position data to correct its output but can display the data for manual computation of the delay.

A. Delay Corrections

The usefulness of potentially highly accurate timing signals relayed through geostationary satellites is complicated by the computation of the propagation delay from the transmitter through the satellite back down to the receiver. Computation of this delay is usually accomplished through the manipulation of orbital elements. From the orbital elements, six constants which describe the satellite's position and velocity at a given instant of time, and a complete description of perturbing forces, it is possible to compute the satellite's position at other times.

Once satellite position is known, the free space propagation delay to any receiver follows directly from simple geometry. The computation of satellite position from orbital elements, however, is complicated and best accomplished using a digital computer.

This type of computation would be difficult for most of the expected users of the GOES time code. However, NBS has developed programmable calculator programs, and a "smart" clock which uses the satellite's position to compute the delay to high accuracy.

B. Programs for Computing Free-Space Path Delay

Several programs to compute the total free space path delay from Wallops Island, Virginia, through the satellite to any point on the earth's surface are available from the Time and Frequency Services Group, 524.06, NBS, Boulder, CO 80303. These programs are written for use with programmable calculators.

10.1.9 SMART CLOCK

A "smart" clock with receiver is shown in figure 10.12. This is essentially an addition

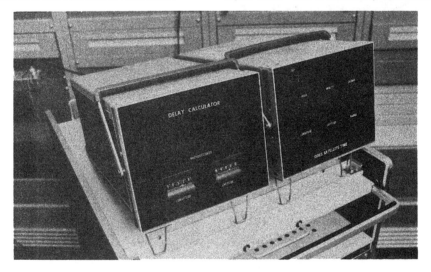

FIGURE 10.12. SMART CLOCK.

182

of a second microprocessor to the "satellite-controlled clock" for the calculation of the free-space propagation delay from the CDA to the clock via the satellite. This delay value is then used with a delay generator to compensate for the free-space path delay.

The "smart" clock uses the same type of four-bit microprocessor as the "satellite-controlled clock." The microprocessor is interfaced to a large scale integration scientific calculator array (math chip) to provide the floating arithmetic and mathematical functions required in the delay calculation. A 1 pps output and satellite position as longitude, latitude, and distance from a reference orbit is obtained continuously from the satellite-controlled clock operating on the satellite's transmitted interrogation channel signal.

User position is entered into the system via thumbwheel switches, and transmitter position (Wallops Island, Virginia) is contained in the microprocessor software. The computed delay drives a programmable delay generator to correct the 1 pps from the satellite controlled clock. The resultant output from the programmable delay generator is a compensated 1 pps, adjusted to be in agreement with the master clock at Wallops Island, which is in turn referenced to UTC(NBS). The hardware is a multiprocessor system consisting of two microprocessors plus a slaved scientific calculator chip and a delay generator.

The delay calculation is dependent on knowledge of the broadcasting satellite's position. This position is predicted in advance by a large scale scientific computer operating on orbital elements obtained from NOAA and sent to Wallops Island via telephone line to be broadcast along with the time code from each satellite. The delay correction system will work with any satellite in a synchronous orbit as long as the satellite's position is known. A calculation is made and the result is latched into the delay generator once per minute. A complete up and down delay calculation requires the execution of about 200 key strokes representing data and mathematical operations and functions under the control of the microprocessor and its associated transistor-transistor-logic (TTL) components.

A block diagram of the smart clock, including receiver and antenna, is shown in figure 10.13. The smart clock has been completely documented (schematics, board layouts, and software)[19].

10.1.10 CLOCK CALIBRATION

To use the GOES signals to calibrate or set another clock, one needs to determine the relationship of the satellite controlled clock's 1 pps relative to UTC(NBS) and to the clock being calibrated. The relationship

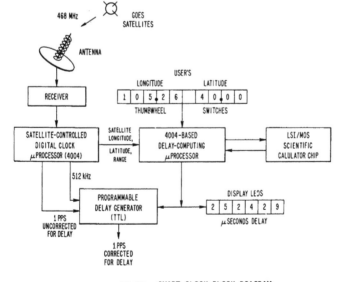

FIGURE 10.13. SMART CLOCK BLOCK DIAGRAM.

183

between the satellite-controlled clock and the clock being calibrated is usually determined by use of a time interval counter. For example, the clock to be calibrated may start the counter, then the satellite controlled clock stops it. The relationship between UTC(NBS) and the user's clock is then known by combining this measured result with the computed difference between the satellite clock's 1 pps and UTC(NBS).

An example of the clock calibration procedure is given below. The clock to be calibrated is located in Boulder, Colorado, where the GOES satellite signals are being received.

Signals received in Boulder (105.26° W Longitude and 40.00° N Latitude) showed the satellite location to be:

134.92° W Longitude

0.38° S Latitude

+46 µs Distance

Using a calculator, the following delays are calculated:

WALLOPS ISLAND TO SATELLITE	133 606 µs
SATELLITE TO BOULDER	127 553 µs
TOTAL PATH DELAY	261 159 µs

To this we must add:

MEASURED EQUIPMENT DELAY	+ 5 192 µs

to get the total delay from Wallops Island to the user's satellite clock 1 pps.

TOTAL DELAY:	266 351 µs

This result can also be written as:

SATELLITE CLOCK - WALLOPS ISLAND CLOCK

= - 266,351 µs,

where the result is negative because the satellite clock is late with respect to the Wallops Island clock (see Chapter 3 for sign conventions).

Since the Wallops Island clock is set to be 260,000 µs early with respect to UTC(NBS):

WALLOPS ISLAND CLOCK - UTC(NBS) = + 260,000 µs

Then, by adding these two equations:

SATELLITE CLOCK - UTC(NBS) = - 6,351 µs   (1)

The measured time interval between the user's clock and the satellite signal is stated as:

USER'S CLOCK - SATELLITE CLOCK = +6,548 µs (2)

Then, by adding equations (1) and (2):

USER'S CLOCK - UTC(NBS) = + 197 µs.   (3)

The smart clock automatically computes the relationship between UTC(NBS) and the output of the satellite-controlled clock and adjusts the difference to be zero; i.e., UTC(NBS) = satellite-controlled clock. The calibration using the smart clock is therefore read directly from a time interval counter.

10.1.11 RESULTS

More than two year's of data from several satellites have been recorded at NBS Boulder using the measurement configuration shown in figure 10.14. The results, on strip chart recorders, have been of the form illustrated in figure 10.15. The phase shift in the uncorrected output is clearly evident. The corrected output is generally a straight line since the satellite position data are updated every four minutes. Remember that the time code message and position data are repeated every half-minute but new position data are input once every 4 minutes.

The equation relating the time recovered from the satellite to the master clock at Wallops Island is given in eq. (4):

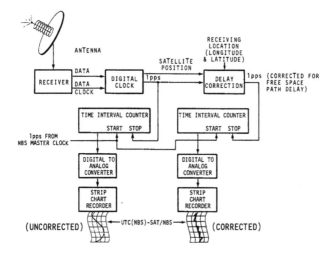

FIGURE 10.14. MEASUREMENT OF GOES SIGNALS.

UTC(NBS) - SAT/USER

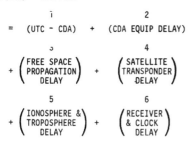

$$= \underset{1}{(UTC - CDA)} + \underset{2}{(CDA\ EQUIP\ DELAY)}$$

$$+ \underset{3}{\begin{pmatrix} FREE\ SPACE \\ PROPAGATION \\ DELAY \end{pmatrix}} + \underset{4}{\begin{pmatrix} SATELLITE \\ TRANSPONDER \\ DELAY \end{pmatrix}}$$

$$+ \underset{5}{\begin{pmatrix} IONOSPHERE\ \& \\ TROPOSPHERE \\ DELAY \end{pmatrix}} + \underset{6}{\begin{pmatrix} RECEIVER \\ \&\ CLOCK \\ DELAY \end{pmatrix}}$$

Term 1 is known to better than 1 microsecond using the data logger which compares the CDA clocks to Loran-C and TV Line-10. This term will be controlled in the next generator system to be installed at Wallops Island to ± 3 microseconds or better. Terms 2, 4, 5 and 6 are expected to remain constants at the few microseconds level and can for the most part be calibrated out of a user's system. Term 3 can be computed as discussed previously, with its accuracy dependent upon the quality of the orbital data given to NBS by NOAA. All measurements to date indicate this term will be accurate to a few tens of microseconds.

Figure 10.16 shows the results of one month's data taken at NBS/Boulder using orbit predictions derived from three sets of orbital elements extrapolating as much as 22 days from their date. The results indicate a system capability of 10 microseconds. Over a period

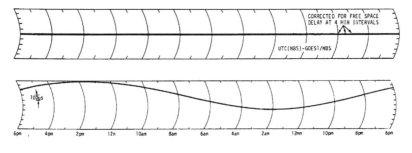

FIGURE 10.15. SATELLITE SIGNALS CORRECTED AND UNCORRECTED.

185

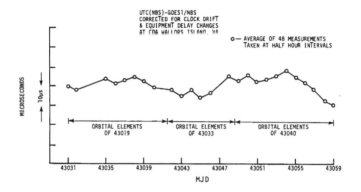

FIGURE 10.16. TYPICAL RESULTS.

of one year, charts have deviated from the "straight line" by as much as 25 microseconds. Using two receivers, there was no correlation in the drift of the receiver and we conclude that the problem resides with the receiving equipment only. A better definition of performance and accuracy cannot be made until data are gathered from widely separated locations in the Western Hemisphere.

10.1.12 PRECAUTIONS

A. Interference

The "land-mobile" services and the GOES interrogation channel use the same frequency allocations (468.8250 and 468.8375 MHz), which means the time code may suffer interference from land-mobile transmissions. This is particularly true in urban areas where there is a high density of land-mobile activity. The satellite frequency allocations are secondary to the land-mobile services. Therefore, any such interference must be accepted by the time signal users. Complaints to the FCC will not result in any adjustments in favor of such users. The spectrum use by satellite and land-mobile is shown in figure 10.17.

Because of the spacing of frequency assignments to the land-mobile users, there is far less interference to the eastern satellite signals than to the western satellite. Therefore, the eastern satellite should be used by those user's situated in large urban areas.

B. Outages

Although the GOES satellites transmit continuously, there may be interruptions during the periods of solar eclipses. The GOES satellites undergo spring and autumn eclipses during a 46-day interval at the vernal and autumnal equinoxes. The eclipses vary from approximately 10 minutes at the beginning and end of eclipse periods to a maximum of approximately 72 minutes at the equinox (fig. 10.18). The eclipses begin 23 days prior to equinox and end 23 days after equinox; i.e., March 1 to April 15 and September 1 to October 15. The outages occur during local midnight for the satellites. In the future, it is expected that GOES will operate through eclipse periods using batteries.

There will also be shutdowns for periodic maintenance at the Wallops Island ground

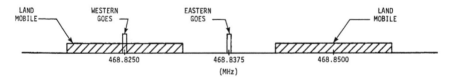

FIGURE 10.17. FREQUENCY USE.

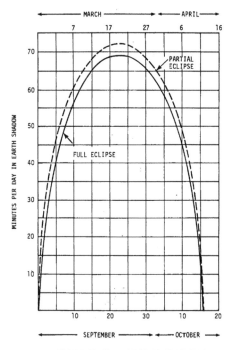

MINUTES PER DAY IN EARTH SHADOW

MARCH ────────► ◄── APRIL ──►
7   17   27   6   16

PARTIAL ECLIPSE

FULL ECLIPSE

10   20   30   10   20

◄─────── SEPTEMBER ───────► ◄── OCTOBER ──►

FIGURE 10.18. SOLAR ECLIPSE TIME

station. These scheduled outages will be reduced in the future as redundancy is added to the Wallops Island facility.

C. Continuity

NBS cannot give an absolute guarantee to the long-term continuance of the GOES time code. The GOES satellites belong to NOAA and are not under the direct control of NBS. However, NBS and NOAA have agreed to include the time code in the GOES satellite transmissions to the maximum extent possible. The GOES system presently has a sufficient number of satellites to operate into the late 1980's, and it is expected that the time code will be included throughout this period.

10.2 THE TRANSIT NAVIGATION SYSTEM

10.2.1 GENERAL

These Navy navigation satellites have been operational since 1964. Known as the

TRANSIT satellites, their design, power level, and coverage provide an excellent source of time and frequency signals for calibration purposes. The signal format used on the system lends itself well to time recovery from which frequency calibrations can be derived using the techniques previously described in this book.

A major difference between TRANSIT and GOES is that TRANSIT, being an orbiting system, experiences a much larger Doppler shift of its radio signal. This is no real problem for the user since the overall system design takes this into account and provides enough information to the users so that accurate time recovery can be achieved.

The biggest difference between GOES and TRANSIT is that the TRANSIT system can only be used intermittently--when a satellite is in view of the receiver.

This system is a good candidate for users who have a worldwide requirement. TRANSIT is, in fact, the only time and frequency system that will work anywhere in the world. GOES is limited at present to the western hemisphere and its oceans.

The satellites broadcast at about 400 MHz. The message sent contains a number of characteristics which allow position fixing and also time recovery. As with the GOES system, TRANSIT allows a user to recover time signals which can be used as a clock or to steer an oscillator's frequency. Unlike GOES, however, the TRANSIT time signal format does not include day-of-the-year information. Accuracy of the overall system is comparable to GOES. Time to better that 10 microseconds and frequency accuracies to 1 part in 10^{10} (for measurement periods of 1 day) can be achieved.

Although TRANSIT is primarily a navigation satellite, the responsible Navy group has recently allowed changes in its operation to provide even more support for its frequency and time function. Coupled with this is the plan to launch additional satellites in the TRANSIT series. Users planning to obtain equipment to use TRANSIT, especially at remote locations, should contact either the U. S. Naval Observatory or the Navy Astronautics Group for the latest information on satellite passes, etc. In addition, manufacturers of TRANSIT equipment are a good source of information.

At this writing, the TRANSIT system is supported by five orbiting satellites. These are in a circular polar orbit at an altitude of about 1100 kilometers. Thus, a user on the earth will see a satellite every 90 minutes or so. In contrast to a GOES-type of transponder,

187

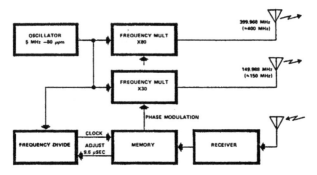

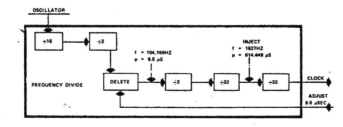

1. SATELLITE OSCILLATOR DESIGN CENTER FREQUENCY = 4,999,600

2. DIVIDER CHAIN RATIO (READ) = 16 x 3 x 2 x 32 x 32 = 98,304

3. 98,304/4,999,600 = 19,662373 ms/BIT

4. 19,662373 x 6103 = 119.9994626 SECONDS

5. 120.0000000 - 119.9994624 = 537.6 MICROSECONDS

6. DIVIDER CHAIN RATIO (DELETE) = 48/4999600 = 9.6 MICROSECONDS

7. 537.6 µs/9.6 µs = 56 CLOCK CORRECTION BITS/2 MIN. INTERVALS

FIGURE 10.19.   TRANSIT SATELLITE CLOCK CONTROL SYSTEM.

TRANSIT has an accurate clock on-board each of its satellites. These are monitored from earth-control stations and corrections are sent to the satellites periodically to keep them on time (see fig. 10.19).

In addition to position information, the satellites send a time mark each two minutes. By monitoring the signals from four ground-monitor stations, the Navy is able to steer the on-board clocks by special commands. For the user, this means that at any time he can be assured that a TRANSIT signal has been carefully controlled for accuracy.

There is much redundancy in the overall system control. New control data are sent to the satellites every 12 hours or so. Almost all the ground equipment used for both navigation and timing contains a fairly

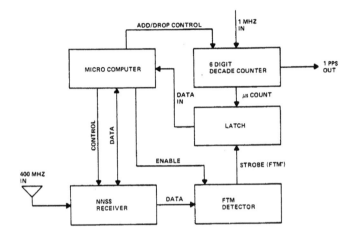

FIGURE 10.20.   TRANSIT TIMING RECEIVER BLOCK DIAGRAM.

sophisticated computer-controlled receiver that allows nearly hands-off operation.

As with the GOES system, the satellite message is designed to allow the receiver to synchronize itself, locate the required bits in the data stream and produce a time mark.

Accuracy of time recovery using TRANSIT depends on several factors. Again, the potential user is encouraged to talk with manufacturers and to contact the USNO, Time Services Division, Washington, DC 20390 before proceeding. At this writing, time recovery accuracy falls within from 10 to 50 microseconds. With suitable care of an operating clock on the ground, this service can provide very valuable calibration information.

10.2.2 EQUIPMENT NEEDED

Equipment costs are higher for TRANSIT than for GOES, but still not unreasonable considering the few alternatives.

Antennas are small, vertical, and omni-directional, making installation simple.

The instrumentation needed by a user will depend on his particular application. In general, he will need a frequency source and clock. Purchase of a receiver for TRANSIT will then allow comparisons to the specifications of the equipment.

It is expected that manufacturers will differ in their implementation of TRANSIT receiving equipment, but a typical requirement is that the user obtain the time of day from another source to within 15 minutes in order to resolve the time ambiguity of TRANSIT. For some applications, it may also be necessary to dial in the user's longitude and latitude.

See figure 10.20 for a typical receiver block diagram.

189

An instrument or device that produces a steady and uniform signal output in whatever form is said to have a frequency of operation. The new quartz crystal wristwatches, for example, vibrate their crystal frequency source at 32,768 times in one second. Most clocks contain an identifiable source of frequency, be it balance wheel, crystal, or pendulum.

Frequency sources are also used in another way. Besides counting their vibrations or oscillations for timekeeping, they are used to set musical instruments and to calibrate many types of electronic devices. A radio dial is marked off in frequency units as are the familiar television channels. The TV channel selector, when set to channel 4, tunes the receiver to obtain a signal at a frequency of about 70,000,000 hertz or 70 MHz.

This chapter deals with the properties of frequency sources and their possible applications to real situations. The idea is that once you understand what the frequency generator will do, you can more intelligently choose the right type of device. Some of the physics of the devices are discussed and finally a table is presented that compares several devices.

The first part of this chapter deals with the use of frequency sources to drive clocks, which is an important application of many high-quality oscillators. There is a strong relationship between time-ordered events and frequency division. Communications systems especially rely on these two means of separating channels of communications. For example, there are a number of stations on the AM radio dial. These are spaced in frequency at about every 10 kHz. Another way to share the spectrum would be to have all the stations at the same frequency, but at different times. The same spectrum is thus time-shared. For obvious reasons of cost and convention, we choose rather to run all the stations at the same time, but on different frequencies.

Common usage allows manufacturers and writers to refer to frequency sources as standards. You will see and hear the term frequency standard mentioned when what is really meant is a frequency source. The word "standard" should properly be reserved for only those frequency generators that serve as references for others. In the U. S. the primary frequency standard resides in Boulder, Colorado, at the National Bureau of Standards Laboratories. But there can be many secondary standards that are calibrated with the single primary standard. Here is where the popular usage of the term "standard" gets misused. People use it to describe high-quality oscillators because their stability is so good that it approaches that of the nation's standard. It is probably not possible to prevent this misuse, so the reader should be warned to note carefully what is meant by the words used to describe the various frequency sources.

11.1 FREQUENCY SOURCES AND CLOCKS

Notice that most clocks--in particular, the very accurate and precise ones--are based on high-quality frequency sources. The reason for this is the intimate relationship between frequency (symbol v, "nu") and time (symbol t). See the definition of frequency in the glossary for more information. If we look at a series of events which are occurring in a somewhat regular fashion; e.g., the rise of the sun every morning, we can state how many of these events occur in a given time period. This number would be the frequency of this series of events. In our example, we could say that the frequency of sunrises is $v = 7$ events per week, or $v = 365$ events per year. "Events per week" or "events per year" would

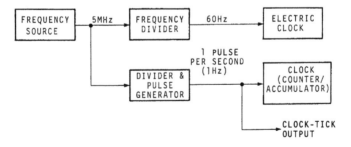

FIGURE 11.1. EXAMPLE OF A CLOCK SYSTEM.

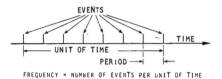

FREQUENCY = NUMBER OF EVENTS PER UNIT OF TIME

$$\text{ACCUMULATED CLOCK TIME} = \frac{\text{TOTAL NUMBER OF EVENTS}}{\text{NUMBER OF EVENTS PER UNIT OF TIME}}$$

POSSIBLE DEFINITION:

UNIT OF TIME = A SPECIFIC NUMBER OF PERIODS OF
A WELL-DEFINED EVENT GENERATOR

FIGURE 11.2. DEFINITION OF TIME AND FREQUENCY.

be called the unit which we used for our frequency number. This frequency number is different for different units. In our example, we assume that we know somehow what a week or a year is; i.e., we relied on some understood definition of our unit of time.

Now, what is the time interval between the events? The answers for our examples are simple: One sunrise succeeds the other after t = 1/7 week, or t = 1/365 year, where we used "week" and "year" as two possible choices for our unit of time.

We have learned two things: (a) for periodic events, the time between the events t is related to the frequency v of their occurrence in the following simple way:

$$v = \frac{1}{t}$$

and (b) that periodic events can be counted to define time; i.e., the generator of the periodic events-the frequency source--can be used as a clock. The frequency becomes a clock by the addition of a counting mechanism for the events.

In the above example, the frequency used is that of the rotating earth. The time between recurring events is one day. The rotating earth has served mankind for thousands of years and remained until very recently the source for the definition of time interval. The counting mechanism which made it a clock was the recording of days and years.

The need to get along for many days without celestial observations, and to measure time intervals which are very much shorter than a day, brought about the invention of clocks. Although there are other types of clocks like the sand clock or the water clock and the sundial, we will discuss only clocks based on electronic frequency generators.

The first clocks based on the pendulum were invented about 400 years ago. This type of clock is still widely used today. The pendulum may be a suspended weight (gravitational pendulum) like in "grandfather" clocks, or the balance wheel (torsion pendulum) of modern wristwatches. The object of our discussion here is today's most advanced oscillators and clocks; however, a close look at traditional clocks will show all the essential features which we will recognize again in our later discussion of quartz crystal and atomic clocks.

The pendulum in our clock is the frequency determining element. In order to have a frequency source, the pendulum has to be set and kept in motion. Thus, a source of energy is necessary together with the means to transfer this energy to the frequency-determining element. In a wristwatch, this source of energy is typically the winding spring or battery. The energy is transferred by a means which is controlled by the pendulum itself (feedback) in order to cause energy transfer in the proper amount at the proper time in synchronism with the movement of the pendulum or balance wheel. We now have a frequency source. The tick motion of a balance wheel could be picked up acoustically, for example, and used as a frequency source. This is actually done by jewelers when they adjust the rate of a clock: the tick-frequency is compared to some (better) frequency source. In order to have a clock, a read-out mechanism is necessary which counts and accumulates the ticks (more accurately, the time interval between the ticks) and displays the result. In our example of a wristwatch, this is accomplished by a suitable set of gears and the moving hands on the clock face or digitally.

We have just discussed how the addition of a mechanism to a frequency source which counts and accumulates and possibly displays the result creates a clock. This task can also be performed by an electronic frequency

192

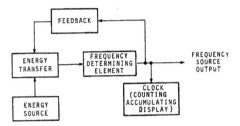

FIGURE 11.3. FREQUENCY SOURCE AND CLOCK.

divider which, for example, derives a frequency of 60 Hz directly from a 5 MHz crystal
oscillator (fig. 11.1). The 60 Hz voltage can
be used to drive an electrical clock motor
similar to those driven by the 60 Hz power
line frequency that we use at home and at
work. Or, an additional electric pulse generator may be used that generates one very
sharp electrical pulse per second. The time
interval of 1 second between the pulses (corresponding to a frequency of 1 Hz) is derived
from the output of our frequency source. The
pulses can be directly used in time comparisons with those of other similar clocks, or
they can drive a counter/accumulator.

The electric power line is a very good
frequency source for many applications. A
clock driven from the power line will keep
excellent time. This is because the power
system is carefully controlled to maintain its
frequency within definite limits. By increasing or decreasing the individual generator
frequencies, the flow of power in the inter-
connected system can be regulated. Each power
company is notified in advance to set their
frequency to a particular value so that the
millions of clocks on the system--those in
homes and offices across the country--will
gain or lose time as required to keep the
clocks correct. The time corrections are
usually done at night.

The basic unit of time is the second
(symbol s). It is defined with reference to a
frequency-determining element. Since 1967, by
international agreement, this element or
"natural pendulum" has been the cesium atom.
The second is defined in the official wording
as "the duration of 9,192,631,770 periods of
the radiation corresponding to the transition
between the two hyperfine levels of the ground
state of the cesium-133 atom." Accordingly,
the frequency of the cesium pendulum is
9,192,631,770 events per second (the cesium
atom is a very rapidly oscillating pendulum).
The unit of frequency is then defined as hertz
(symbol Hz) which means the repetitive occurrence of one event per second (the use of
"hertz" is preferred to the older term "cycle
per second", cps).

11.2 THE PERFORMANCE OF FREQUENCY SOURCES

The performance of a frequency source is
usually described in terms of its accuracy,
reproducibility, and stability. These are
defined as:

ACCURACY:

How well does it relate to the definition? In the case of frequency, this means
how well it relates to the definition of a
second (in terms of cesium) mentioned
previously.

REPRODUCIBILITY:

If you built a number of such frequency
sources and adjusted them, how well would they
agree in frequency? This term obviously
applies to sources that are manufactured and
then tested to see how they differ.

STABILITY:

Once the device is set to a given frequency, how well does it generate that value
during some period of time?

It is obvious from these definitions that
frequency accuracy will be of great interest
in scientific measurements and in the evaluation and intercomparison of the most advanced
devices. Good reproducibility is an asset in
applications where it is important to expect
some agreement among several devices. However, the stability of a frequency source is
usually most important to the user.

The frequency stability (symbol σ,
"sigma") of a frequency source will depend on
many things that might cause frequency
changes. Frequency stability can be measured
by taking a reasonably large number of successive readings on an electronic counter which
counts the frequency of the device to be
evaluated. Each counter reading (in hertz) is
obtained by counting the output frequency for
some specified time (the sampling time "tau",
symbol τ). This sampling time can usually be
chosen by simply adjusting a knob on the
counter; for example, a sampling time of 0.1
s or 1 s or 10 s may be chosen. The result
can be expected to change as τ changes. So
the stability needs to be defined for a
certain "tau".

Further, it might be expected that variations in the readings of measured frequency
would tend to average out if observed long
enough. This is not always so. The stability
of the frequency source (sigma or σ) will
therefore usually depend on the sampling time
of the measurement and tends to get smaller
with longer sampling times, τ. But, again,
there are many exceptions to this.

193

It may happen that the frequency fluctuations at some later time are partially caused by, or depend to some degree upon, the previous fluctuations. In this case, the computed value of σ will depend on the particular way in which the many counter readings are averaged and evaluated. Another influence on measured stability depends on whether the counter starts counting again immediately after completion of the preceeding count or if some time elapses ("dead-time") before counting starts again.

Finally, the electronic circuits used in measuring frequency stability will·have a finite response time; i.e., they cannot follow frequency fluctuations faster than some given rate. For example, our eyes cannot register light fluctuations which occur faster than about every 1/10 of a second; using $v = 1/t$, we say that the eye has a frequency response of 10 Hz, or that its bandwidth is only 10 Hz; i.e., the eye cannot follow frequencies higher than 10 Hz.

In order to measure frequency stabilities for sampling times smaller than some value τ, our measurement equipment has to provide for an electronic frequency bandwidth which is larger than 1/τ.

To summarize: A recommended way of properly measuring and describing frequency stability is the following: (a) make sure that the frequency bandwidth of the total measuring setup is larger than 1/τ, where τ is the smallest desired sampling time; (b) use a counter with a deadtime as small as possible (the dead-time should be less than the reciprocal bandwidth; if not, use the computation procedures that exist to allow for larger deadtimes); (c) take a sufficiently large number of readings at a given sampling time which is held constant and compute:

$$\sigma = \sqrt{\frac{\text{ADDITION OF THE SQUARES OF THE DIFFERENCES BETWEEN SUCCESSIVE READINGS}}{2 \times \text{TOTAL NUMBER OF DIFFERENCES USED}}}$$

(In the scientific literature this σ is often called the square root of the Allan variance.) The counter readings can be taken in Hz; σ will then have the dimensions of Hz; (d) repeat step (c) for other sampling times τ and tabulate or plot σ as it depends on τ.

Commonly, σ will be expressed as a relative value. Existing literature often uses the term "fractional" instead of relative, so the reader should treat the two terms equally; they mean the same thing. The value obtained for the frequency stability is divided by the carrier frequency. For example, if a

frequency stability of $\sigma_{\delta v} = 10$ Hz were measured at a carrier frequency of $v = 5$ MHz (MHz = megahertz = million Hz) then the fractional or relative frequency stability would be

$$\sigma_y = \sigma_{\delta v/v} = 2 \times 10^{-6}.$$

We denote the kind of σ we are using by a subscript, δv ("delta nu"), referring to frequency fluctuation (measured in Hz) or δv/v = y, referring to relative frequency fluctuations (dimensionless). A stability of one part in a million is thus

$$\sigma_y = 1 \times 10^{-6},$$

and one part in a trillion is written as

$$\sigma_y = 1 \times 10^{-12}.$$

There is a sound basis for using the relative frequency stability σ_y instead of using frequency stability $\sigma_{\delta v}$ directly. σ_y is a numeric and is independent of the actual operating frequency of the oscillator being discussed. Thus, it is possible to compare the stability of a 10 MHz oscillator with one that has an output at 10 kHz.

In fact, many high-quality oscillators have outputs at 1, 5, and 10 MHz. But an application for these sources might require the generation of a different frequency, e.g., the television color subcarrier signal at 3.58 MHz. If the original source, before the synthesis of the TV signal, had a relative frequency stability of one part per million (1 x 10-6), then the 3.58 MHz signal would have the same relative frequency stability σ_y.

This assumes, of course, that the synthesis or generation of new frequencies does not change the stability of the frequency source. This is generally true for most applications.

11.3 USING RELATIVE FREQUENCY STABILITY DATA

Suppose you have a clock. Let's say its frequency source is a crystal oscillator, and you want to know the time error. If you know enough about the oscillator, you can compute the time error. You can even predict the error ahead of time--fairly accurately, too.

First, let's define a few things that we will need to write an equation. There are only about three things that can happen to our clock--assuming we don't drop it or pull the plug!

1. The clock could be reading the wrong time because we didn't set it at all or because we set it wrong. Call this the initial time error Δt_0. Until we correct it, we have to have it in our equation as a constant.

2. The frequency of the oscillator driving the clock may be fluctuating. If that is the case, then we will have a term in our equation that relates the fluctuation and the length of time it has been drifting. Let's call this length of time t. It would be the time from zero time (when the clock was right or we measured the error Δt_0) till now. From what we said about the stability, we see that we need a term such as:

$$\sigma_y(\tau) \times t.$$

Actually, we know the value of τ that we want. It's simply t, the time since we last set the clock. So our term becomes approximately:

$$\sigma_y(\tau = t) \times t.$$

This then takes care of the instabilities of the oscillator frequency driving our clock.

3. There is one more possible error. Was the oscillator "on frequency" when it started to drift? If it was not, then we must account for this offset. To do this, we define the relative frequency offset as:

$$\frac{\Delta v}{v} \times t.$$

This term means that our oscillator may have had a frequency error when it started to drift. In other words, we allow for an arbitrary starting point.

To review then, we have an oscillator connected to a clock and it may or may not have had a time error Δt_0 in addition to having or not having some drift $\sigma_y \times t$ from either the correct or incorrect frequency $\Delta v/v \times t$.

Addition of these time errors gives us the clock error at time t:

$$\Delta t = \Delta t_0 + \sigma_y(\tau = t) \times t + \frac{\Delta v}{v} \times t.$$

The equation shows how the relative frequency offset and stability accumulate time error as t grows larger. Let's plug in some numbers and see what happens to our clock. Assume the oscillator was off frequency by 1 part in 10^{11}, fluctuates by 1 part in 10^{11} per day, and had a 1 microsecond error yesterday at this time. Reading the graph we see that our clock error would be about three microseconds.

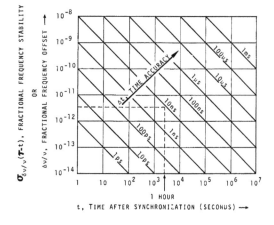

FIGURE 11.4. RELATIONSHIPS BETWEEN CLOCK ACCURACY, FREQUENCY STABILITY, AND FREQUENCY OFFSET.

195

We get almost 1 microsecond error from fluctuations, 1 microsecond from frequency offset and one from our initial setting error.

11.4 RESONATORS

The performance of an oscillator depends a lot on its basic frequency-determining element. This element determines the frequency by its resonance behavior. Thus, it is called a resonator. Some examples of resonators are a rod clamped only at one end that can vibrate; a block of solid material that can contract and expand and thus vibrate; an electrical capacitor-coil combination (tank circuit) in which the energy can oscillate back and forth; and a coil in which an electric current can create a magnetic field which can oscillate between its two possible polarities (a magnetic dipole). These devices have one thing in common--they will vibrate or oscillate if they are excited. The method of excitation may be a mechanical pulse, an electrical pulse, or a sudden surge of an electric or

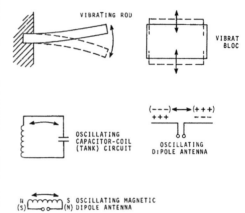

FIGURE 11.5. EXAMPLES OF RESONATORS.

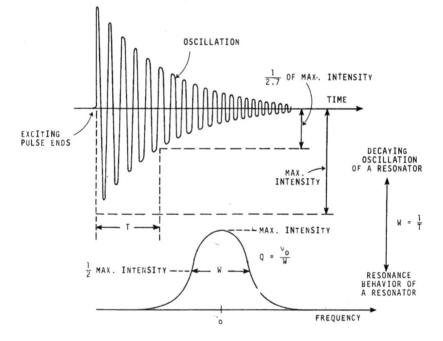

FIGURE 11.6. DECAY TIME, LINEWIDTH, AND Q-VALUE OF A RESONATOR.

196

magnetic field. These devices exhibit resonance. They are resonators with a well-defined center frequency which is a characteristic of the length of the rod, the thickness of the block, the size of the capacitor and coil, or the length of the antennas. Once excited, the oscillations will gradually die out. The decay time is determined by the losses in the resonator. Some of these losses are internal friction, or electrical resistance. In any case, the oscillation energy is finally transformed into heat. If there were no losses, the oscillations would never stop and we would have the ideal resonator. The greater the losses, the faster the oscillations will decay. We can use the decay time of the oscillations (symbol T) to describe the quality of our resonator. The larger T is, the better the resonator. "Better" for us means that the device favors one frequency over all others.

An alternate way of measuring resonance behavior is to use a separate oscillator, couple it to the resonator, and slowly sweep the frequency of the oscillator. We will find a signal frequency at which the resonator will oscillate, with maximum intensity. Above or below this resonance frequency, the response of the resonator will lessen until it ceases to respond. This frequency interval around resonance, for which the resonator response is relatively strong, is called the resonance linewidth (symbol W). There is a simple relationship between W and T, the decay time:

$$W = \frac{1}{T} \text{ Hz.}$$

This equation is an approximation and applies only to certain types of resonance. Another factor would have to be inserted if the resonance curve were different from that shown in figure 11.6. However, this factor is nearly equal to one: T is actually defined as the decay of the amplitude to 37% of its original value: 37% is $1/e$, e is the base of natural logarithms and equals 2.71828.

As we did in the case of σ (page 194), it is convenient to consider the linewidth in a fractional way. The fractional linewidth would be W/v_0 where the symbol v_0 is used for the resonant frequency. A more popular term is the "quality factor" (Q) of the device. Q is defined as the reciprocal fractional linewidth

$$Q \equiv \frac{v_0}{W}$$

For accurate frequency sources and good clocks, we want to have elements with very high Q's. A Q value of one million (10^6) is very common. For example, a 1 MHz crystal oscillator with a Q of a million would have a linewidth of 1 hertz. The oscillator would then favor operation at a frequency within 1 hertz of a megahertz. If this oscillator were measured, its frequency should be within its one hertz range, i.e., within a part per million. In fact, it can be made much better than that. Parts per billion are typical.

In summary then, we want high-quality resonators if our end product is to have good accuracy and stability characteristics. The discussion that follows in this section deals with several types of high-quality frequency sources.

Many kinds of frequency-determining elements have been and are being used in frequency sources. They can be grouped into three classes:

> mechanical resonators
>
> electronic resonators
>
> atomic resonators

As far as mechanical resonators are concerned, we will only discuss one kind, the quartz crystal. Other mechanical resonators like the pendulum and the tuning fork are of less importance in today's high performance frequency sources, although they have been historically very important and are still widely used in watches. For similar reasons, we will also omit the discussion of electronic resonators such as LC tank circuits and microwave cavities. Atomic resonators form the heart of our most accurate frequency sources and clocks, and will therefore be extensively discussed.

11.5 PRIMARY AND SECONDARY STANDARDS

At this point, we should briefly discuss the terms "primary frequency standard" and "secondary frequency standard". These terms refer to the application of the devices; any frequency source, regardless of its accuracy or stability, can be a primary frequency standard, if it is used as the sole calibration reference for other frequency sources. A secondary frequency standard is a device which is occasionally calibrated against a primary frequency standard but operationally serves as the working reference for other frequency sources.

The machines used for primary standards, like those at NBS, Boulder, NRC, Canada, and in Germany and England, are in a class by themselves. Because of the way they have been built and operated, they can be evaluated. This means that experimental data is taken and

FIGURE 11.7. THE U.S. PRIMARY FREQUENCY STANDARD, NBS-6.

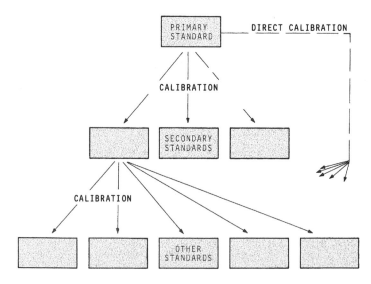

FIGURE 11.8. HIERARCHY OF FREQUENCY STANDARDS.

used to calculate the errors in their output frequency due to all known causes. It is because of this careful evaluation that the accuracy can be stated without having to check these machines against others. This concept will be discussed in detail in connection with the cesium beam frequency sources.

11.6 QUARTZ CRYSTAL OSCILLATORS

Quartz crystal oscillators are widely used. They range from the tiny units found in wristwatches to elaborate instruments in laboratories. Sometimes they are called crystal oscillators and sometimes quartz oscillators. Whatever the name, they dominate the field of good frequency sources.

The quartz crystal in the oscillator is a mechanical resonator. The resonator's oscillations have to be excited and sensed. This is done by taking advantage of the piezoelectric effect in the quartz crystal. The piezoelectric effect (figure 11.9) results when mechanical compression of the crystal generates a voltage across the crystal. Conversely, the application of an external voltage across the crystal causes the crystal to expand or contract depending on the polarity of the voltage. A crystal is not a homogeneous medium but has a certain preferred direction; thus, the piezoelectric effect has a direc-

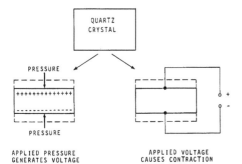

APPLIED PRESSURE
GENERATES VOLTAGE

APPLIED VOLTAGE
CAUSES CONTRACTION

FIGURE 11.9. THE PIEZOELECTRIC EFFECT.

tional dependence with respect to the orientation of the crystal. In order to take advantage of the piezoelectric effect, one has to cut a crystal resonator from the original crystal block in a well-defined way with respect to the crystallographic directions. The raw material used today is natural or synthetic quartz. A crystal is cut out of the raw crystal material in the desired orienta-

tion with the aid of optical techniques which allow the determination of the crystallographic axes. The high-precision, final orientation of the cut and the tuning to the desired frequency is then done by grinding and etching, controlled by x-ray methods.

The quartz crystal can be cut and electrically excited in a variety of ways. The most common types of vibrations (modes) are the longitudinal and thickness modes, the flexure (bending) mode, the torsional mode, and the shear mode. In order to use the piezoelectric effect, metal electrodes have to be attached to the crystal surfaces so that the desired mode is excited.

The electrodes are typically created as extremely thin metallic coatings by vacuum evaporation of metals. Electric leads are attached to the electrodes by soldering. They also serve as the mounting support, thus freely suspending the quartz crystal. In order to least upset the mechanical vibrations of the crystal, the electrode-support leads are attached at points where no vibrational motion occurs (nodes). The crystal is then packaged and the case is either filled with a protective gas or else evacuated.

Using the crystal, an oscillator can be built by adding an electronic amplifier feedback and a power supply. The oscillator output frequency is determined by the quartz crystal resonator. Its frequency in turn is determined by the physical dimensions of the crystal together with the properties of the crystalline quartz used. The actual operating frequency thus depends on the orientation of the cut, the particular mode, and the dimensions of the crystal. As an example, we find that the resonance frequency, for a longitudinal mode of vibration, is approximately:

$$v_0 = 2.7 \times 10^3 \times \frac{1}{L},$$

where L is the length of the crystal. If L is in meters, the resonance frequency v_0 will be in hertz. The equation allows us to estimate the size of the crystals. For example, a 100 kHz crystal will have a length of just a few centimeters. We see that the manufacture of quartz crystals with resonance frequencies above 10 MHz is hardly possible. However, we can excite resonators not only in their so-called fundamental mode (previously discussed), but also at multiples (overtones) of this fundamental resonance frequency. An example of this is the violin string which also can be caused to vibrate at frequencies which are multiples of its fundamental. Quartz crystals which are designed for excitation at multiples of their fundamental resonance are called overtone crystals.

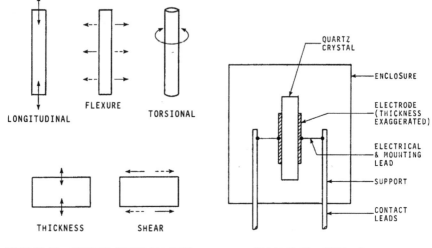

LONGITUDINAL   FLEXURE   TORSIONAL

THICKNESS   SHEAR

FIGURE 11.10.   PRINCIPAL VIBRATIONAL MODES
                OF QUARTZ CRYSTALS.

FIGURE 11.11.   TYPICAL QUARTZ
                CRYSTAL MOUNT.

QUARTZ
CRYSTAL

ENCLOSURE

ELECTRODE
(THICKNESS
EXAGGERATED)

ELECTRICAL
& MOUNTING
LEAD

SUPPORT

CONTACT
LEADS

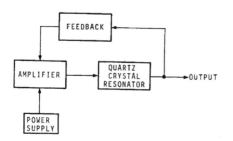

FIGURE 11.12.   QUARTZ CRYSTAL
                OSCILLATOR.

FEEDBACK

AMPLIFIER

QUARTZ
CRYSTAL
RESONATOR

OUTPUT

POWER
SUPPLY

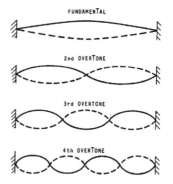

FUNDAMENTAL

2nd OVERTONE

3rd OVERTONE

4th OVERTONE

FIGURE 11.13.   FUNDAMENTAL AND OVERTONE
                RESONANCE FREQUENCIES.

200

11.6.1 TEMPERATURE AND AGING OF CRYSTALS

Temperature and aging are two effects, among others, which are important in the design of crystals and crystal oscillators. There is a temperature dependence of the quartz crystal that affects resonance frequency, and there is also a drift of the resonance frequency as time goes on.

The temperature dependence is caused by a slight change in the elastic properties of the crystal. Certain cuts (crystallographic orientations of the crystal) minimize this effect over a rather wide range of temperatures, most notably the so-called "AT" and "GT" cuts. Temperature coefficients of less than one part in 100 million per degree (celsius or kelvin) are possible. In other words, the fractional frequency change will be less than 10^{-8} with one degree of temperature change. This small effect requires care in the design of a crystal oscillator if very high frequency stabilities over longer times (hours or days) are desired. If large environmental temperature fluctuations are to be tolerated, the crystals themselves are enclosed in electronically regulated ovens which maintain a constant temperature. In certain crystal oscillators this is done to better than 1/1000 of a degree.

An alternate solution to the temperature problem is the so-called temperature-compensated crystal oscillator or TCXO. An additional frequency-determining element in the oscillator, which can be just a small capacitor, allows us to tune the oscillator over a limited range. If a temperature sensor is added to cause a change in this capacitor, the change in resonance frequency of the crystal resonator can be cancelled. Capacitors whose values change with an applied voltage (varactors) are used. The applied voltage is derived from a temperature-sensing circuit.

The TCXO does not necessarily require further temperature control by an oven. However, there is a drawback to this approach. By adding another frequency-determining element, the crystal resonator loses a corresponding part of its control on the output frequency of the whole oscillator. The long-term stability (days) of TCXO's is therefore below that of crystals with a good oven control. TCXO's are used in small, usually portable units of relatively lower performance. They are used where frequency stabilities from day to day and frequency changes (over some tens of degrees of temperature) of not better than 10^{-9} are needed.

Drift, or aging, is a common behavior of all crystal oscillators. It is a nearly linear (uniform) change in resonance frequency with time. Drift often is negative, meaning that the resonance frequency decreases. A frequency decrease could be interpreted as an increase in the crystal size. Many things have been considered as the cause: contamination (depositing of foreign material) on the surfaces; changes in the electrodes or the metallic plating; reforming of loose (from grinding and etching) surface material; or changes in the internal crystal structure. All of these are possibly caused or enhanced by the vibrating motion of the oscillating crystal. Recent crystal holder design improvements, combined with clean vacuum enclosures, have led to a reduction of aging to about 10^{-11} per day. For a 5 MHz crystal with a thickness of a little less than a millimeter, this aging corresponds to an absolute thickness change of only 10^{-11} of a millimeter, or less than 0.1% of the diameter of an atom.

When a crystal oscillator is first turned on, it will not usually oscillate at its original frequency. It will exhibit a "warm-up" due to temperature stabilization of the crystal resonator and its oven. For some time then, often many days, a large but diminishing drift occurs until it reaches its operating point. Its frequency then might also be substantially different (as compared to its stability and aging performance) from its frequency before power was turned off.

11.6.2 QUARTZ CRYSTAL OSCILLATOR PERFORMANCE

Crystal resonators have Q-values which are typically in the range from $Q = 10^5$ to almost 10^7. These are very high Q-values as compared to most other resonators except atomic resonators.

These high Q-values are an essential prerequisite for the excellent stability performance of crystal oscillators. The best available oscillators have stabilities of a few parts in 10^{13} for sampling times from one second to a day. There is some experimental evidence that some crystal resonators may, in the future, perform even better! The limitations are primarily caused by noise from electronic components in the oscillator circuits. This noise may possibly be reduced by selection of low-noise components--transistors, capacitors, etc., and by special circuit designs. Thus, there is a reasonable chance that crystal oscillator stability can reach values of better than 10^{-13} for sampling times of seconds to hours. For times shorter than one second, stability is often determined by additive noise in the output amplifiers. This can be reduced by a crystal filter in the output. The long-term stability beyond several hours sampling time is determined by

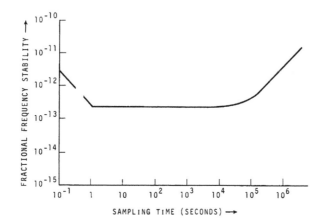

FIGURE 11.14. FREQUENCY STABILITY OF THE BETTER QUARTZ CRYSTAL OSCILLATORS.

aging and by external influences such as line voltage variations, temperature fluctuations, etc. A specification of accuracy is not easily stated for crystal oscillators. Without frequency calibration, they can be manufactured by controlling the dimensions of the quartz used to within 10^{-6} of a nominal frequency. If they are calibrated against a high accuracy frequency source, they maintain this calibration (accuracy) in accordance with their long-term stability characteristics. A crystal with a low aging characteristic of 10^{-11} per day will maintain its accuracy to a few parts in 10^8 for a year without recalibration.

It is apparent, therefore, that crystal oscillators require calibration at least once a year. They may need it more often depending on the application. Frequency adjustments are made with a small added capacitor in much the same way as was discussed in connection with the TCXO. The most stable crystal oscillators, with the lowest aging rate, cost approximately $2,000. They have a volume of a few thousand cubic centimeters, and require input power of about 10 watts. They have an elaborate crystal oven for temperature control, well-designed electronics, and usually several output frequencies which are derived from the oscillator with frequency dividers and multipliers. These high performance devices use 2.5 or 5 MHz crystal resonators having Q-values of a few million.

Cheaper and smaller crystal oscillators are available in a variety of designs with corresponding reductions in frequency stability and/or environmental insensitivity. Costs can go down to below $100, sizes to a few cubic centimeters, and power requirements to

less than 0.1 watt. The reliability of crystal oscillators is usually not limited by the crystal itself, but also depends on the accompanying electronic circuits.

11.7 ATOMIC RESONANCE DEVICES

Commercially available atomic frequency oscillators are based on resonances in atoms at microwave frequencies--in the range from 1 to 100 GHz (gigahertz or billion hertz). We may picture atomic resonance as a magnetic dipole antenna. We typically deal with a great many of these dipole antennas (atomic resonators), and we can, as a simplification, separate them into two kinds: Receiving antennas receive energy from a magnetic field that oscillates at their resonance frequency, much like a TV antenna at home. Transmitting antennas radiate energy at their resonance frequency just like the transmitting antenna of a broadcast station. A physicist might say that the atoms that act like a receiving antenna are in the "lower state" and those acting like a transmitting antenna are in the "upper state."

> There is a peculiarity with atoms: an atom changes from the upper to the lower state upon emission of a well-defined amount of energy. Correspondingly, an atom changes from the lower to the upper state after receiving an equal amount of energy at the atomic resonance frequency.

If we have a great many atoms as in a gas, we will find that the number of upper

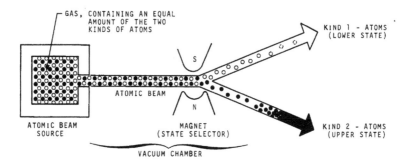

GAS, CONTAINING AN EQUAL AMOUNT OF THE TWO KINDS OF ATOMS

KIND 1 - ATOMS (LOWER STATE)

S

ATOMIC BEAM

N

KIND 2 - ATOMS (UPPER STATE)

ATOMIC BEAM SOURCE

MAGNET (STATE SELECTOR)

VACUUM CHAMBER

FIGURE 11.15. SPATIAL STATE SELECTION.

state atoms is nearly equal to those in a lower state. This has an important consequence: If this gas is placed in a magnetic field that oscillates at the atomic resonance frequency, all atoms will resonate. Nearly half of the atoms receive (absorb) energy from the field, and the other half emit (add) energy of an equivalent amount to the field. The net effect is almost zero. The gas as a whole acts as if it has almost no resonance even though the individual atoms do resonate. From this we see that in order to achieve atomic resonance, we have to somehow change the relative amounts of the two kinds of atoms. The upper or the lower state has to be in the majority. The way in which this is done determines the design of our atomic frequency oscillator.

11.7.1 STATE SELECTION

There are two important methods used to accomplish what is called state selection--the change in the relative numbers of the two kinds of atoms (upper state and lower state):

Spatial state selection relies on an actual sorting procedure where the two atomic states are sorted into different directions in space. One of the states can then be used; the other is discarded. An actual system may produce an atomic gas by heating some substance in an oven. The atoms leave the oven through a hole and form an atomic beam in an adjacent vacuum chamber. The atomic beam is then passed through a rather strong magnet which causes the separation of the beam into two beams, one containing the upper state atoms, the other the lower state atoms. Recall that our picture of an atom is that of a magnetic dipole antenna. The magnet exerts a force on these magnetic dipoles. The force acts in opposite directions for the two different atomic states (we note this without further explanation).

Optical state selection takes advantage of the fact that atoms have more than just one kind of resonance. Other resonances typically correspond to infrared or visible (light) frequencies. We can excite one of these resonances by shining intense light of the

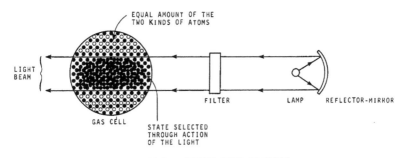

EQUAL AMOUNT OF THE TWO KINDS OF ATOMS

LIGHT BEAM

FILTER   LAMP   REFLECTOR-MIRROR

GAS CELL

STATE SELECTED THROUGH ACTION OF THE LIGHT

FIGURE 11.16. OPTICAL STATE SELECTION.

203

right frequency on the atoms. If the light is filtered carefully, very monochromatic--one color (a very well-defined frequency)--light is obtained. If the light frequency is chosen properly, only one kind of atom will resonate. It will be found that this light frequency is different from the corresponding light resonance of the other kind of atom. Light resonance can thus "remove" one kind of atom and provide for the desired majority of the other kind.

11.7.2 HOW TO DETECT RESONANCE

We are now almost in a position to assemble our atomic frequency oscillator. We still need some means to detect atomic resonance. A short waveguide or microwave cavity is typically used. Such a cavity has microwave resonances which are determined by its physical size. The electrical losses of this cavity are determined by the electrical conductivity of the cavity material. A cavity may be shaped like a cylinder or a box. It resonates in a way quite similar to an organ pipe in the case of sound waves. The best known example of a microwave cavity is the microwave oven where food is cooked by placing it into the resonance field of a cavity.

In order to observe atomic resonance, we have to place the state-selected atoms inside this cavity and subject them, for some specified time, to a microwave radio signal at their resonance frequency. The microwave signal will change the relative number of atoms in the two states. If all atoms were initially in the upper state, we will find some in the lower state after the microwave signal has acted upon them for some time. If the frequency of the microwave signal equals the atomic resonance frequency, the transfer of atoms from one state to the other reaches a maximum. The center of the atomic resonance is thus found by monitoring the number of atoms in one of the two states while varying the microwave frequency until the number of atoms in the lower state reaches a maximum, or the number of atoms in the upper state reaches a minimum. An electric signal can be obtained from the state selector which is fed back to the crystal oscillator generating the microwave signal. Thus, an automatic servo can be built that keeps the crystal oscillator locked to the atomic resonance.

Detecting the effect of the microwave signal on the atoms can be done in three ways. Two of these relate to the methods of state selection.

In <u>atom detection</u>, the atoms which leave the cavity as an atomic beam are passed through the field of a magnet that spatially sorts the two states. An atom detector is placed to intercept one of these states. The output of the detector indicates the number of atoms in the upper (or lower) state.

With <u>optical detection</u>, the atoms are optically state selected. A photo-detector for light can be placed in a position such that the light which has interacted with the atoms is detected. Since the light removes atoms from one of the states, its intensity

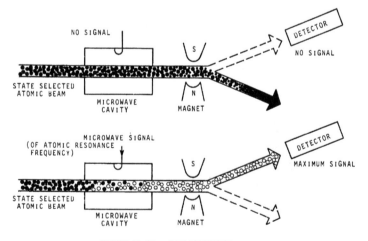

FIGURE 11.17. ATOM DETECTION.

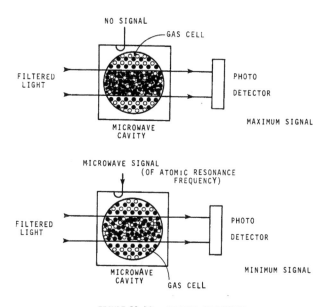

NO SIGNAL

GAS CELL

FILTERED LIGHT

MICROWAVE CAVITY

PHOTO DETECTOR

MAXIMUM SIGNAL

MICROWAVE SIGNAL
(OF ATOMIC RESONANCE FREQUENCY)

FILTERED LIGHT

PHOTO DETECTOR

MINIMUM SIGNAL

MICROWAVE CAVITY

GAS CELL

FIGURE 11.18. OPTICAL DETECTION.

will change if atoms that resonate at the light frequency are added. Such an addition takes place as a result of the microwave signal. The microwave, in effect, transfers atoms from a state with no light interaction to another which takes part in the light resonance. The light at the photo-detector, therefore, is a measure of the number of atoms in one of the states.

For microwave detection, the microwave radio signal is transmitted through the cavity to a microwave detector. The atoms will either add energy to the signal if they are initially in the upper state or subtract energy if they are initially in the lower state. The microwave power level at the detector is thus a measure of the number of atoms changing state.

11.7.3 ATOMIC OSCILLATORS

The frequency-determining element is the atomic resonator which consists of a cavity containing the state-selected atoms and some means of detecting the change in the number of atoms in the two states. A control signal, related to the number of atoms that changed their state due to the action of the microwave signal, is fed back to the crystal oscillator. The oscillator and the associated frequency multiplier provide the energy-transferring

means. As discussed before, good crystal oscillators are available at frequencies of several megahertz. The atomic resonances are at gigahertz frequencies. The crystal oscillator frequency therefore has to be multiplied by a factor of about 1000. Some power supply has to provide the energy necessary to drive the oscillator, multiplier, and possibly the atom state selector.

Very high Q values can be achieved with atomic resonators. Recall that resonator Q is related to the time T, which describes the average duration of oscillations of the resonator. In atomic resonators, we find two chief causes for the dying out of oscillations. First is the collision of the resonating atoms with each other and with surrounding walls. Each collision usually terminates the oscillation. The second is rather obvious. The atoms may simply leave the region of microwave signal interaction. In an atomic beam machine, the atoms enter the cavity, traverse the cavity in a certain time, and then leave the cavity. As an example, let us assume that we have a cavity of one meter length which is traversed by atoms with an average speed of 100 meters per second (a typical value). We assume an atomic resonance frequency of 10 GHz. The interaction time T is then 1/100 second, the linewidth is 100 Hz, and the Q value is 10^8. This Q value is considerably better than that of a quartz oscillator.

205

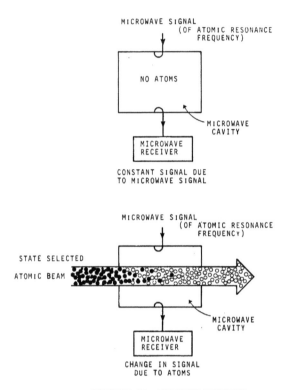

CONSTANT SIGNAL DUE
TO MICROWAVE SIGNAL

CHANGE IN SIGNAL
DUE TO ATOMS

FIGURE 11.19. MICROWAVE DETECTION.

11.7.4 ATOMIC RESONATOR FREQUENCY STABILITY AND ACCURACY

Before we discuss what might cause changes in the output frequency of an atomic oscillator, we should make a very important distinction between quartz crystal and atomic oscillators: The atomic resonance frequency itself is given to us by the nature of our device. Unlike the quartz crystal, it will not drift or age. Thus, atomic resonators with Q values of 10^8 or higher may be expected to have accuracies of one part in 10^8 or better. This is because we will not be able to shift the frequency further away from resonance than the linewidth.

The following listing is not exhaustive but gives only the minor perturbing effects on atomic resonators:

Random noise in the crystal oscillator, the detector, the microwave cavity, and the

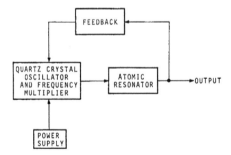

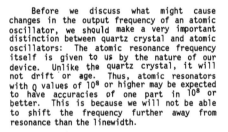

FIGURE 11.20. ATOMIC FREQUENCY OSCILLATOR.

206

frequency synthesizer will cause corresponding changes of the output frequency of the atomic oscillator.

The microwave cavity is itself a resonator. Thus, we have an additional frequency-determining element in addition to the atoms. The cavity influences the output frequency by pulling the combined resonance frequency to a value which usually lies between the resonances of the atom and cavity. This necessitates the tuning of the cavity itself so it is equal to the atomic resonance. The requirements for tuning are relaxed if the Q value of the atomic resonance is as high as possible and the cavity Q as low as possible.

The microwave signal will never be just one single frequency but will have a certain distribution of frequencies or spectrum. If the exciting microwave signal has an unsymmetrical distribution of frequencies, a frequency pulling occurs which is much like the mechanism of cavity pulling. This effect can be made negligible, however, by careful electronic design.

Collisions between the atoms and between the atoms and the walls of a vessel (gas cell) in which the atoms may be contained will not only shorten the duration of the oscillation but also cause a frequency shift. Obviously, these effects can be minimized by having low atom densities and no walls, if possible.

An apparent change in frequency when the source moves relative to an observer is called Doppler effect. Everyone experiences this by listening to an approaching train whistle or automobile sounds. Here, the moving objects are the oscillating atoms and the "observer" is the microwave cavity. The Doppler effect can be greatly reduced by choosing a particular direction of the atomic beam with respect to the direction of the oscillating magnetic microwave field in the cavity. In our acoustic example, this corresponds to the passing of the whistle at some large distance instead of having the whistle move directly towards the observer. An alternate way of reducing the Doppler effect is the containment of atoms in a region which is small compared to the wavelength of the microwave radiation; e.g., by placing a gas cell inside of the cavity.

Of the several effects which are discussed here, magnetic field change is the only one that directly affects atomic resonance. Recall that we used a small magnetic dipole antenna as the model for the atomic resonance. If this antenna is placed in an external static magnetic field (as that of a permanent magnet), the tuning of the antenna changes; that is, the atomic resonance frequency changes. This necessitates magnetic shield-ing, which is a characteristic design feature of all presently used atomic frequency oscillators. The shielding is usually quite elaborate and reduces the effect of magnetic fields, such as the earth's magnetic field, to 1% or less of their external value. This residual magnetic field can be measured quite precisely by using the atomic resonance itself. In fact, the overall measurement precision is so good that magnetic field effects do not seriously limit the accuracy of atomic frequency oscillators. However, changes in the external magnetic field, or the movement of the device through a varying field, as in a relocation, may perturb the frequency of the atomic oscillator.

Generally speaking, one tries to minimize all of the effects listed above and to keep their influence as stable with time as possible. Such an approach suffices for most applications. But the magnetic effect is an exception. The residual magnetic field inside the shields must be evaluated and possibly reset any time after changes in the external field occur, say, after moving the device. The magnetic shields may also have to be demagnetized. For laboratory standards, however, where frequency accuracy is the primary purpose, all effects must be evaluated in a series of experiments, and reevaluations must be done occasionally in order to detect changes.

11.8 AVAILABLE ATOMIC FREQUENCY DEVICES

We will now discuss the design and performance of the three types of atomic frequency oscillators which are currently in operational use.

11.8.1 CESIUM BEAM FREQUENCY OSCILLATORS

For a cesium oscillator, atomic resonance is at 9,192,631,770 Hz. The oscillator is based on the atomic beam method using spatial state selection and atom detection. An oven contains the cesium metal. If heated to about 100°C, enough cesium gas will be produced to form an atomic beam which leaves the oven through one or many channels into a vacuum chamber. This chamber is evacuated to less than 10^{-9} of atmospheric pressure. The beam traverses first the state-selecting magnet, then the microwave cavity. Typically, a cavity with separated interrogation regions is used. This design offers certain advantages over a simple cavity of a length equivalent to the separation of the two regions. In the cavity, an external microwave signal acts on the beam. The beam finally reaches the atom detector after passing another state-selecting magnet. The atom detector is simply a

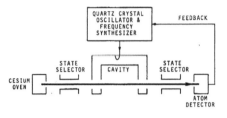

FIGURE 11.21.   CESIUM BEAM FREQUENCY
OSCILLATOR.

tungsten wire which is heated to about 900°C by passing an electric current through it. This wire is biased with a few volts dc and cesium atoms which hit it become electrically charged or ionized and can be collected on an auxiliary electrode. The stream of electrically-charged atoms at this electrode represents an electric current which is amplified, detected, and used in the feedback network.

The speed of the atoms and the length of the cavity determine the Q value of the atomic resonator. Typical atom speeds are 100 meters per second. In non-laboratory devices, which have to be reasonably small, the cavity is about 0.2 meter long. The corresponding interaction time T is two-thousandths of a second. We can calculate the linewidth as a few hundred hertz and a Q value of 10^7 (see section 11.4). In laboratory devices, one can

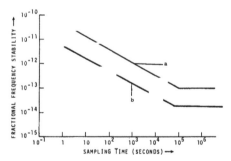

a)   Typical performance.

b)   Typical performance of recently available high beam intensity tubes.

Individual units may perform slightly worse or better than shown. Modern laboratory standards perform still better.

FIGURE 11.22.   FREQUENCY STABILITY OF
COMMERCIAL CESIUM BEAM
FREQUENCY OSCILLATORS.

use very long cavities. Cavities of up to 4 meters long are used, leading to Q values of 10^8. The fractional frequency stability of laboratory and commercial devices can reach parts in 10^{14} at sampling times of less than one hour to days. The short-term frequency stability is limited by fluctuations in the atomic beam intensity, "shot noise," which is basic and unavoidable. These fluctuations affect the frequency stability less as more intense atomic beams are used. This approach, which is becoming available in both commercial and laboratory devices, improves the stability. In contrast to commercial devices, the laboratory oscillators are designed to allow a more complete and easier evaluation of all effects on the frequency. Cesium oscillators are used extensively where high reproducibility and long-term stability are needed for sampling times of more than a day. For most applications, cesium oscillators need not be calibrated. They are the work horses in most of today's accurate frequency and time distribution services.

11.8.2   RUBIDIUM GAS CELL
FREQUENCY OSCILLATORS

Rubidium atomic resonance is at 6,834,682,608 Hz. The oscillator is based on the gas cell method using optical state selection and optical detection. The gas cell contains rubidium gas at about 10^{-9} of atmospheric pressure. In order to reduce the effect of collisions among the rubidium atoms, argon, an inert buffer gas, is introduced into the cell at about 1/1000 of atmospheric pressure. This allows lifetimes of the rubidium atom oscillations of about 1/100 second. The oscillation lifetime T is still limited by atom collisions. We can calculate a corresponding linewidth of about 100 Hz and a Q value of 10^7. Atomic collisions as well as the simultaneous action of the light and the microwave signals on the same atom cause frequency shifts of the order of 10^{-9}. These

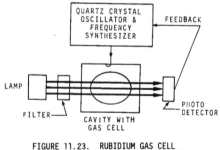

FIGURE 11.23.   RUBIDIUM GAS CELL
FREQUENCY OSCILLATOR.

frequency shifts depend strongly on the composition, temperature, and pressure of the buffer gas and on the intensity of the light. As a result, rubidium gas cells vary in their resonance frequency by as much as 10^{-9}, depending on the particular setting of the frequency shifting parameters during manufacture. Since these influences cannot be expected to remain unchanged with time, rubidium oscillators need initial calibration and also recalibration because they exhibit a frequency drift or aging like crystal oscillators. The stability performance of rubidium oscillators is nevertheless quite spectacular. At one-second sampling times, they display a stability of better than 10^{-11} and perform near the 10^{-13} level for sampling times of up to a day. For longer averaging times, the frequency stability is affected by the frequency drift which is typically 1 part in 10^{11} per month. This is much less than the drift of crystal oscillators. Commercial devices with the above-mentioned performance are available as well as rubidium oscillators of lower performance at significantly reduced costs.

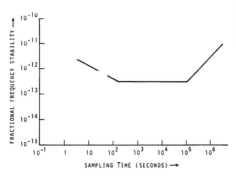

FIGURE 11.24. FREQUENCY STABILITY OF COMMERCIAL RUBIDIUM GAS CELL FREQUENCY OSCILLATORS.

Rubidium oscillators are used wherever excellent medium-term--minutes to a day--stability is needed. They are used where reduced costs, size, and weight, as compared to cesium oscillators, are important. Rubidium devices are used where a crystal oscillator, with its more frequent needs for recalibration and its greater environmental sensitivity, would not suffice.

11.8.3 ATOMIC HYDROGEN MASERS

Maser is an acronym meaning "microwave amplification by stimulated emission of radiation." The atomic resonance frequency is at 1,420,405,752 Hz. The oscillator is based on the atomic beam method using spatial state selection and microwave detection. All natural hydrogen gas is composed of hydrogen molecules. Each hydrogen molecule is formed by chemical bonding of two hydrogen atoms. The beam source is a radio frequency gas discharge in molecular hydrogen. This discharge produces atomic hydrogen with high efficiency. The atomic hydrogen beam leaves the source through one or many channels into a vacuum chamber. The beam then traverses a state-selecting magnet and enters a storage bulb in the microwave cavity. The storage bulb is made from quartz glass which has low electric losses and thus does not significantly alter the cavity Q. The storage bulb is evacuated to less than 10^{-11} of atmospheric pressure. Its inner walls are lined with Teflon. This coating allows many collisions of the hydrogen atoms with the walls without significantly disturbing the oscillations of the atoms. The underlying physical mechanisms are not yet fully understood. The storage bulb is typically 0.15 meter in diameter and dimensioned in such a way as to hold hydrogen atoms for about one second.

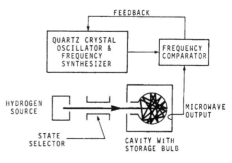

FIGURE 11.25. HYDROGEN MASER OSCILLATOR.

After about one second, the atoms leave the bulb and thus also leave the microwave cavity. We can calculate a linewidth of about 1 Hz and a Q value of about 10^9. This is the highest Q value in all presently-used atomic standards. Higher Q values are realized in experimental devices such as methane-stabilized helium-neon lasers and ion storage devices as well as in microwave cavities which

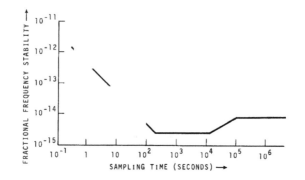

FIGURE 11.26. FREQUENCY STABILITY OF A HYDROGEN MASER OSCILLATOR.
The curve is a composite of the best documented per-
formances of several differently designed devices.
However, there is sufficient evidence to indicate
that a device could be built that performs as indicated. Programs are underway twoard this goal.

are cooled to the superconducting state. If
the intensity of the hydrogen beam, which
consists only of upper state atoms (emitting
atoms), is sufficiently large and if the
cavity losses are sufficiently low, self-
oscillation will start in the cavity. The
maser itself will generate a microwave signal.
We then have a maser-oscillator with an output
frequency directly derived from the atomic
resonance. A crystal oscillator can be locked
to this frequency by frequency comparison
techniques. When compared to a cesium oscillator, the hydrogen maser is not quite as
accurate. This is because of experimental
difficulties in the evaluation of the frequency shift due to the collisions of the
hydrogen atoms with the Teflon surface of
the storage bulb. In order to obtain self-
oscillation, the cavity Q must be relatively
high. Thus, cavity pulling is relatively
strong, and it limits the long-term stability
over periods longer than several days. Values
of long-term stability are not better than
those of cesium oscillators. However, for
periods of a few seconds to a day, the hydrogen maser has the best stability of all
existing oscillators. Its application is
rather limited to uses where these stabilities
are critical and where a rather bulky device

is not a handicap. Unlike cesium and rubidium
oscillators, hydrogen masers have not yet been
evaluated under adverse environmental conditions. The number of hydrogen masers in use
is very small compared to the numbers of
cesium beam and rubidium devices.

11.9 TRENDS

All devices which were discussed in this
report have improved over the past years in
one or more of the following aspects: reliability, size, weight, stability performance,
and environmental insensitivity. A considerable amount of effort is still being expended
to go further in this direction; however, it
would be inappropriate here to try to estimate
the improvement of performance specifications
which might (or might not) be achieved in the
future.

There are several other devices, designs,
and concepts which have been studied or are
being investigated with some potential for
future use in frequency standards. However,
they will not be discussed here.

TABLE 11.1   COMPARISON OF FREQUENCY SOURCES.

LISTED ARE TYPICAL CHARACTERISTICS OF FREQUENCY SOURCES (1975). THE CRYSTAL OSCILLATORS RANGE FROM SMALL PLUG-INS TO THE BEST LABORATORY UNITS. VALUES ARE GIVEN FOR THE VERY SMALL CESIUM AND RUBIDIUM OSCILLATORS AS WELL AS THE LARGER STANDARD UNITS. THE HYDROGEN MASER IS SHOWN FOR COMPARISON SINCE VERY FEW HAVE BEEN SOLD COMMERCIALLY.

CHARACTERISTICS	KIND OF OSCILLATOR			
	CRYSTAL	CESIUM	RUBIDIUM GAS CELL	HYDROGEN MASER
BASIC RESONATOR FREQUENCY	10 kHz TO 100 MHz	9,192,631,770 Hz	6,834,682,608 Hz	1,420,405,752 Hz
OUTPUT FREQUENCIES PROVIDED	10 kHz TO 100 MHz	1, 5, 10 MHz TYPICAL	1, 5, 10 MHz TYPICAL	1, 5, 10 MHz TYPICAL
RESONATOR Q	10^4 TO 10^6	10^7 TO 10^8	10^7	10^9
RELATIVE FREQUENCY STABILITY, SHORT-TERM, 1 SECOND	10^{-6} TO 10^{-12}	5×10^{-11} TO 5×10^{-13}	2×10^{-11} TO 5×10^{-12}	5×10^{-13}
RELATIVE FREQUENCY STABILITY, LONG-TERM, 1 DAY	10^{-6} TO 10^{-12}	10^{-13} TO 10^{-14}	5×10^{-12} TO 3×10^{-13}	10^{-13} TO 10^{-14}
PRINCIPAL CAUSES OF LONG-TERM INSTABILITY	AGING OF CRYSTAL, AGING OF ELECTRONIC COMPONENTS, ENVIRON-MENTAL EFFECTS	COMPONENT AGING	LIGHT SOURCE AGING, FILTER & GAS CELL AGING, ENVIRON-MENTAL EFFECTS	CAVITY PULLING, ENVIRONMENTAL EFFECTS
TIME FOR CLOCK TO BE IN ERROR 1 MICROSECOND	1 s TO 10 DAYS	1 WEEK TO 1 MONTH	1 TO 10 DAYS	1 WEEK TO 1 MONTH
FRACTIONAL FREQUENCY REPRODUCIBILITY	NOT APPLICABLE. MUST CALIBRATE.	1×10^{-11} TO 2×10^{-12}	1×10^{-10}	5×10^{-13}
FRACTIONAL FREQUENCY DRIFT	1×10^{-9} TO 1×10^{-11} PER DAY	$< 5 \times 10^{-13}$ PER YEAR	1×10^{-11} PER MONTH	$< 5 \times 10^{-13}$ PER YEAR
PRINCIPAL ENVIRONMENTAL EFFECTS	MOTION, TEMPERATURE, CRYSTAL DRIVE LEVEL	MAGNETIC FIELD, ACCELERATIONS, TEMPERATURE CHANGE	MAGNETIC FIELD, TEMPERATURE CHANGE, ATMOS-PHERIC PRESSURE	MAGNETIC FIELD, TEMPERATURE CHANGE
START-UP AFTER BEING OFF	SECONDS TO HOURS (MAY HAVE SYSTEM-ATIC OFFSET)	30 TO 60 MINUTES	10 TO 60 MINUTES	FEW HOURS
RESONATOR RELIABILITY, TIME BETWEEN REPLACEMENTS	VERY RELIABLE	3 YEARS	3 YEARS	NO DATA
TYPICAL SIZE, CUBIC CENTIMETERS	10 TO 10,000	10,000 TO 700,000	1,000 TO 20,000	300,000
TYPICAL WEIGHT IN KILOGRAMS (POUNDS)	0.1 TO 10 (2 OZ TO 22 LB)	16 TO 400 (35 TO 70 LB)	1 TO 20 (3 TO 45 LB)	200 (460 LB)
POWER CONSUMED, WATTS	0.1 TO 15	30 TO 200	12 TO 35	40 TO 100
ESTIMATED PRICE	$100 TO $5000	$15,000 TO $20,000	$3,000 TO $10,000	$100,000 TO $200,000

CHAPTER 12. <u>SUMMARY OF AVAILABLE SERVICES</u>

In the previous chapters, we discussed the various time and frequency dissemination services that can be used for calibrations. The following tables should serve as a quick reference to these services. Table 12.1 summarizes the major characteristics of the different services. Tables 12.2 and 12.3 show the different means of calibrating time and frequency sources and the accuracies that can be obtained.

TABLE 12.1. CHARACTERISTICS OF THE MAJOR T&F DISSEMINATION SYSTEMS

DISSEMINATION TECHNIQUES		ACCURACY FREQUENCY SYNCHRONI-ZATION	ACCURACY FOR DATE TRANSFER	AMBIGUITY	COVERAGE FOR STATED ACCURACY
VLF RADIO	GBR, NBA, OMEGA, ETC.	1 x 10-11	ENVELOPE 1 - 10 ms	1 CYCLE	NEARLY GLOBAL
LF RADIO	STANDARD FREQUENCY BROADCAST (e.g., WWVB)	1×10^{-11} PHASE 24h	ENVELOPE 1 - 10 ms	YEAR	USA - LIMITED (WWVB)
	LORAN-C	5×10^{-12}	.1 μs (GND) 50 μs (SKY)	TOC 15 MIN PHASE 10 μs	SPECIAL AREAS
HF/MF RADIO	STANDARD FREQUENCY BROADCAST (e.g., WWV)	1×10^{-7}	1000 μs	CODE - YEAR VOICE - 1 DAY TICK - 1 s	HEMISPHERE
TELE-VISION (VHF/SHF RADIO	PASSIVE LINE-10	1×10^{-11}	1 μs	33 ms	NETWORK COVERAGE
	COLOR SUBCARRIER	1×10^{-11}	NA	NA	
SATEL-LITE (UHF RADIO)	GOES	3×10^{-10}	30 μs	1 YEAR	WESTERN HEMISPHERE
	TRANSIT	3×10^{-10}	30 μs	15 MINS	
PORTABLE CLOCKS	PHYSICAL TRANSFER	1×10^{-13}	100 ns	N/A	LIMITED BY TRANSPORTATION

213

TABLE 12.2. HOW FREQUENCY IS CALIBRATED

COMMERCIALLY AVAILABLE FREQUENCY SOURCE	ACCURACY NEEDED (RELATIVE)	METHOD OF CALIBRATION
TUNING FORKS	$1 \times 10{-2}$ TO 1×10^{-3}	COMMERCIAL FREQUENCY COUNTERS; STANDARD T&F BROADCASTS
ELECTRIC POWER GENERATORS	$\cong 2 \times 10^{-5}$ (1×10^{-3} Hz IN 60 Hz)	COMMERCIAL FREQUENCY COUNTERS; STANDARD T&F BROADCASTS
QUARTZ OSCILLATORS	1×10^{-3} TO $! \times 10^{-10}$	COMMERCIAL FREQUENCY COUNTER UP TO 1×10-7; FREE-RUNNING ATOMIC OSCILLATORS OR STANDARD T&F BROADCASTS FOR 1×10-7 OR BETTER
ATOMIC OSCILLATORS	1×10-9 TO $\cong 1 \times 10$-13	COMMERCIAL CESIUM STANDARDS, HYDROGEN MASERS AND NATIONAL PRIMARY STANDARDS. RUBIDIUM OSCILLATORS CAN BE CALIBRATED ON SITE WITH A COMMERCIAL CESIUM STANDARD OR A HYDROGEN MASER TO AN ACCURACY OF ABOUT 5×10^{-12}. FOR CALIBRATION AT A DISTANCE, USE THESE SAME SOURCES AND LINE-OF-SIGHT MICROWAVE OR AN LF OR VLF BROADCAST. FOR ACCURACIES IN THE RANGE OF 5×10^{-12} TO 1×10^{-13}, COMPARE TO A NATIONAL PRIMARY STANDARD AND USE ON-SITE CALIBRATION OR LF OR VLF BROADCASTS.

TABLE 12.3. HOW TIME IS CALIBRATED

SOURCE OF TIME	ACCURACY NEEDED	METHOD OF CALIBRATION
REGULAR (ESCAPEMENT) WATCHES AND CLOCKS	10 SEC $\cong 1 \times 10^{-4}$/DAY (10 TIMES BETTER THAN USUAL WATCH)	TELEPHONE TIME FROM NBS OR CERTAIN SELECT CITIES; STANDARD RADIO BROADCASTS
ELECTRIC CLOCKS	1 SEC $\cong 1 \times 10^{-5}$/DAY (MORE ACCURATE THAN MOST POWER COMPANIES CAN HOLD)	TELEPHONE TIME FROM NBS OR CERTAIN SELECT CITIES; STANDARD RADIO BROADCASTS
QUARTZ WATCHES AND CLOCKS	0.1 SEC $\cong 1 \times 10^{-6}$/DAY (STABILITY THIS GOOD BUT CAN'T BE SET THIS WELL)	TELEPHONE TIME FROM NBS OR CERTAIN SELECT CITIES; STANDARD RADIO BROADCASTS
PRECISION QUARTZ CLOCKS	1 μs $\cong 1 \times 10^{-10}$/DAY	CESIUM CLOCK: STANDARD RADIO BROADCASTS
ATOMIC CLOCKS	FROM $\cong 3 \times 10^{-3}$ SEC TO $\cong 3 \times 10$-6 SEC ON YEARLY BASIS (i.e., 1×10-10 TO 1×10-13)	NBS OR USNO TIME SCALES VIA PORTABLE CLOCK OR LF OR VLF BROADCASTS

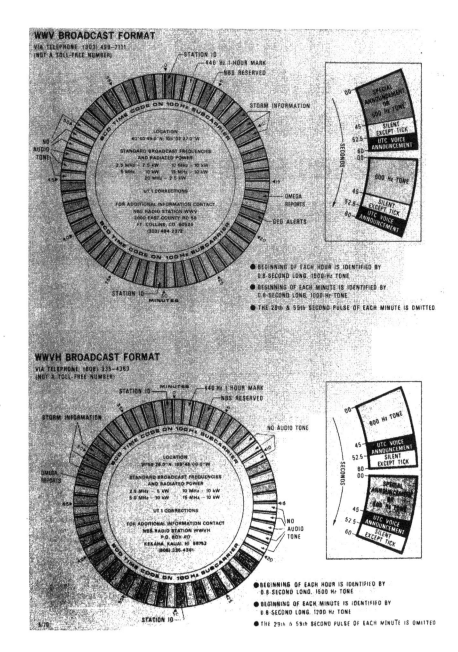

FIGURE 12.1.   WWV/WWVH HOURLY BROADCAST FORMAT.

215

[1] Barnes, J. A., "A Non-mathematical Discussion of Some Basic Concepts of Precise Time Measurement," On Frequency (published by Tracor, Inc.), Vol. II, No. II (May 1971).

[2] Blair, Byron E., ed., Time and Frequency: Theory and Fundamentals, Nat. Bur. Stand. Monograph 140 (May 1974). (Order from: Superintendent of Documents, Government Printing Office, Washington, DC 20402, SD Catalog No. C13.44:140.)

[3] Bagley, A., "Frequency and Time Measurements," Handbuch der Physik, Vol. XXIII, (Springer-Verlag Co., Berlin, 1966).

[4] Czech, J., Oscilloscope Measuring Technique: Principles and Applications of Modern Cathode Ray Oscilloscopes (Springer-Verlag Co., New York, 1965).

[5] Evans, J., "Time and Time Measurement," Electronics World, pp. 25-29 (July 1966).

[6] Willrodt, M. J., "Frequency and Frequency Measurement," Electronics World, pp. 25-28 (October 1966).

[7] Hironaka, N. and C. Trembath, "The Use of National Bureau of Standards High Frequency Broadcasts for Time and Frequency Calibrations," Nat. Bur. Stand. Tech. Note 668 (May 1975).

[8] Viezbicke, P. P., "NBS Frequency-Time Broadcast Station WWV, Fort Collins, Colorado," Nat. Bur. Stand. Tech. Note 611 (October 1971).

[9] Howe, Sandra L., "NBS Time and Frequency Dissemination Services," Nat. Bur. Stand. Spec. Publ. 432 (January 1976).

[10] Milton, J. B., "Standard Time and Frequency: Its Generation, Control, and Dissemination by the National Bureau of Standards," Nat. Bur. Stand. Tech. Note 656 (June 1974).

[11] Tolman, J., V. Ptacek, A. Soucek and R. Stecher, "Microsecond Clock Comparison by Means of TV Synchronizing Pulses," IEEE Trans. Instr. and Meas., Vol. IM-16, pp. 247-254 (September 1967).

[12] Soucek, A., "Travel Time Stability on TV Relay Links," Radio and Television, No. 5, pp. 29-31 (1969).

[13] Davis, D. D., J. L. Jespersen and G. Kamas, "The Use of Television Signals for Time and Frequency Dissemination," Proc. IEEE (Letter), Vol. 58, No. 6, pp. 931-933 (June 1970).

[14] Davis, D. D., "Transmission of Time/Frequency Signals in the Vertical Interval," Synopses of Papers (108th Tech. Conf. & Equip. Exhibit, Soc. Mot. Pict. & Tele. Eng., New York, NY, October 4-9, 1970), Paper No. 43 (October 1970).

[15] Davis, D. D., "Frequency Standard Hides in Every Color TV Set," Electronics (May 10, 1971).

[16] Davis, D. D., "Calibrating Crystal Oscillators with TV Color-reference Signals," Electronics (March 1975).

[17] Davis, D. D., "A Microprocessor Data Logging System for Utilizing TV as a Time/Frequency Transfer Standard.," (Proc. 8th Annual Precise Time and Time Interval (PTTI) Applications and Planning Meeting, Goddard Space Flight Center, Greenbelt, MD, Nov. 30 - Dec. 2, 1976, pp. 167-181).

[18] Hamilton, W. F. and D. W. Hanson, "A Synchronous Satellite Time Delay Computer," Nat. Bur. Stand. Tech. Note 638 (July 1973). (Order from U.S. Government Printing Office, Superintendent of Documents, Washington, DC 20402. SD Catalog No. C13.46:638. Price: $.45.)

[19] Cateora, J. V., D. W. Hanson and D. D. Davis, "Automatic Path Delay Corrections to GOES Satellite Time Broadcasts," Nat. Bur. Stand. Tech. Note 1003 (February 1978).

[20] Cateora, J. V., D. D. Davis and D. W. Hanson, "A Satellite-Controlled Digital Clock," Nat. Bur. Stand. Tech. Note 681 (June 1976).

[21] Smith, H. M., "International Time and Frequency Coordination," Proc. IEEE, Vol. 60 (1972).

[22] Howe, D. A., "Frequency Domain Stability Measurements: A Tutorial Introduction," Nat. Bur. Stand. Tech. Note 679 (1976).

[23] Allan, D. W., "The Measurement of Frequency and Frequency Stability of Precision Oscillators," Nat. Bur. Stand. Tech. Note 669 (1975).

[24] Barnes, J. A., et al, "Characterization of Frequency Stability," IEEE Trans. Instr. and Meas., Vol. IM-20 (1971).

[25] Allan, D. W., H. Hellwig and D. J. Glaze, "An Accuracy Algorithm for an Atomic Time Scale," Metrologia, Vol. 11 (1975).

[26] Siegmann, "An Introduction to Lasers and Masers," Chapter 2, McGraw-Hill (1971).

[27] Gerber, E. A. and R. A. Sykes, "A Quarter Century of Progress in the Theory and Development of Crystals for Frequency Control and Selection," Proc. 25th Annual Symp. on Frequency Control, Ft. Monmouth, NJ (1971).

[28] Ramsey, N. F., "History of Atomic and Molecular Control of Frequency and Time," Proc. 25th Annual Symp. on Frequency Control, Ft. Monmouth, NJ (1971).

[29] Audoin, C. and J. Vanier, "Atomic Frequency Standards and Clocks," J. Phys. E: Scientific Instruments, Vol. 9 (1976).

[30] Hellwig, H., "Clocks and Measurements of Time and Frequency," WESCON Technical Papers, Session 32 (1976).

The following definitions are offered as an aid to the reader. However, they may not conform to the definitions as recommended by the CCIR or other appropriate organizations. Further, we have not attempted to include all possible unfamiliar terms. Many terms are fully explained in the text and have not been repeated here.

ACCURACY - how well something agrees with the definition or standard. In the case of frequency, an oscillator is said to be accurate if it agrees with the master oscillator at the National Bureau of Standards in Boulder, Colorado. There is often confusion about the definition of the terms resolution, precision, and accuracy.

AGING - the process whereby crystals change their characteristics. A quartz crystal of 100 kHz may age until its frequency becomes 100.01 kHz. The process of aging is not well understood.

AMBIGUITY - the properties of something that allows it to have more than one possible meaning. In the case of a clock, if it displays 3 hours, 5 minutes, it could be morning or night. The clock is then said to be ambiguous to the hour. Ambiguity is overcome in time codes by sending the day number. This makes that particular time code ambiguous to the year.

ANALOG FREQUENCY METERS - that class of instruments that uses needle movements rather than lighted digits to indicate frequency. Accuracies obtained with these meters is often less than with their digital counterparts.

AT CUTS - one of the many ways in which bulk quartz material can be sliced to produce frequency controlling crystals. The AT cut is usually used for frequencies above 500 kHz and has excellent frequency versus temperature characteristics.

ATOM DETECTION - a process used in atomic clocks to detect atoms.

ATOMIC TIME (TA) - the time obtained by counting cycles of an atomic clock as opposed to time based on the earth's rotation. Atomic time is extremely uniform.

AUTOMATIC GAIN CONTROL (AGC) - a circuit used in communications receivers so that both weak and strong stations can be heard with nominally equal volume.

BALANCE WHEEL - a small wheel in watches under control of the hairspring. The oscillating balance wheel determines the rate of the watch.

BURST - the name given to a short signal consisting of a number of cycles. The WWV tick could properly be called a burst.

CALIBRATION - a frequency source can be calibrated by measuring its frequency difference with respect to the National Bureau of Standards frequency standard. It is also possible to calibrate one frequency source against another.

CESIUM STANDARD (Cs) - This term is often used to describe any oscillator that uses cesium to obtain atomic resonance.

CHRONOMETER - a very accurate clock or watch, usually used on ships for navigation. This name has also been applied to digital instruments driven from quartz crystals.

COLOR BAR COMPARATOR - an instrument that is used to compare a user's oscillator to the television color subcarrier by putting a colored vertical bar on the television screen.

CYCLE CORRECTION - the correction applied to low frequency radio data that allows for the chance that the receiver may be locked to the incorrect carrier cycle. In the case of Loran-C, one cycle is 10 microseconds. For the WWV tone, one cycle would be 1 millisecond.

DATE - a complete description of the time of an event. The date should properly include year, day, hour, minute, and second. It can be given by day number in a particular year and also by Julian Day.

DECADE COUNTING ASSEMBLY (DCA) - a circuit used in counters that counts from 0 to 9 before it overflows. This enables the counter to display units of tens, hundreds, etc.

DIGITAL PHASE SHIFTER - an instrument using digital techniques which causes a phase shift to occur in a signal. In the case of 1 MHz, for example, phase shifts usually occur in one-tenth microsecond steps or less.

DIGITAL SLEWING SYSTEM - a means whereby a continuous phase shift in a signal can be caused by digital control.

DIURNAL PHASE SHIFT - the phase shift (diurnal means daily) associated with sunrise or sunset on low frequency radio paths.

DOPPLER EFFECT - an apparent change in frequency caused by motion of either the transmitter or receiver. In the case of radio waves, a Doppler effect occurs when the ionosphere changes its position.

DOUBLE CONVERSION - the conversion in two steps from the radio carrier frequency to the final intermediate frequency. High quality radio receivers use double conversion to avoid reception of image frequencies.

DRIFT (FREQUENCY DRIFT) - frequency drift is the change in frequency of a frequency source--caused by temperature or aging, for example.

DUT1 - the number to be added to UTC to obtain an approximation of a UT1 date. The value of DUT1 is given on the U. S., Canadian, and most other similar broadcasts so that a listener can correct the time as heard to make it agree with UT1 within 0.1 second.

EMISSION DELAY - a delay deliberately inserted at the transmitter to time the various signals being emitted. In Loran-C, it allows a number of stations (a chain) to transmit on the same frequency.

EPHEMERIS TIME (ET) - an astronomical time system based on the orbital motion of the earth. See also Universal Time.

EPOCH - a date selected as a point of reference, usually in astronomy. This word is very often used instead of the term date or time, but should be discouraged because of ambiguities in its meanings.

FADING - the characteristic of radio waves which causes their received amplitude to decrease. Fading is caused by a combination of several radio waves that are not in exact phase.

FERRITE LOOP - a loop antenna that uses a rod made of iron powder. It is commonly used in small AM broadcast receivers. It can also be used to receive WWVB at 60 kHz.

FLYWHEEL - an oscillator or frequency source used to keep a clock going when its usual source of frequency is lost. Many digital clocks use an RC oscillator to "flywheel" when the 60 hertz power is disconnected.

FREQUENCY - if T is the period of a repetitive phenomena, then the frequency is f = 1/T. In SI units, the period is expressed in seconds and the frequency is in hertz. Note that different symbols can be used (especially in mathematics) to identify a particular frequency. In this book, both "f" and "n" have been used (see Chapters 3 and 11). The reader should take note of what is meant in each case.

FREQUENCY DISCRIMINATOR - an instrument or circuit which can distinguish frequency changes. FM radio circuits use discriminators and they are also used to measure frequency differences.

FREQUENCY - the name given to an instrument developed by NBS to measure frequency by use of the color television subcarrier.

FREQUENCY-SHIFT KEYING (FSK) - a means of modulating a radio carrier by changing its frequency by a small amount.

FREQUENCY SYNTHESIZERS - instruments or circuits that change one input frequency to one or more different output frequencies. A communications system, for example, could use a 5 MHz quartz crystal oscillator and synthesize the carrier or data frequencies needed for communications.

GREENWICH MEAN TIME (GMT) - related to solar time at Greenwich, England, which has been used extensively in the past. At this writing, Greenwich Mean Time can be considered equivalent to Coordinated Universal Time (UTC) which is broadcast from all standard time and frequency radio stations.

220

GROUP REPETITION RATE - the rate of recurrence of specified groups of pulses. In the case of Loran-C, the group repetition rate is altered between different chains to avoid interference.

GT CUTS - refers to the direction in which a quartz crystal blank is cut. GT crystals have excellent frequency stability but are usually larger in physical size than the AT cuts.

HARMONIC GENERATOR - an instrument or circuit that generates harmonics of the input signal. A harmonic generator provides a means of using a lower frequency source to calibrate high frequency signal generators.

HETERODYNE - the means by which new frequencies are generated by mixing two signals in a nonlinear device such as a vacuum tube, transistor, or diode mixer. For example, a superheterodyne receiver converts all incoming frequencies by heterodyne action to a common intermediate frequency.

HETERODYNE CONVERTER - a frequency converter that uses the heterodyne process to convert input signals to a range where the instruments can measure their frequency.

HYSTERESIS - a lag effect in a body when the force acting on it is changed.

IMAGE REJECTION - the ability of a superheterodyne receiver to reject an unwanted signal which is just as far on one side of its local oscillator as the desired frequency is on the other side.

INTERMEDIATE FREQUENCY (IF) - the frequency in a superheterodyne receiver to which allE incoming frequencies are converted. The intermediate frequency is usually a low frequency somewhere between 450 and 3000 kHz.

IONOSPHERE - the outer part of the earth's atmosphere. The ionosphere consists of a series of constantly changing layers of ionized molecules. Many radio waves reflect back to earth from the ionosphere.

JITTER - usually used to describe the small changes in a signal with time. Phase jitter

JITTER - usually used to describe the small changes in a signal with time. Phase jitter causes a counter to be triggered either early or late.

JULIAN DAY (JD) - the Julian day number is obtained by counting days with a starting point of noon January 1, 4713 B.C., which was Julian Day zero. This is one way of telling what day it is with the least possible ambiguity.

LEADING EDGE - on a narrow pulse like that from a clock, it is the first-occurring change and can be either negative or positive-going.

LEAP SECOND - a one-second interval either inserted or deleted in a clock to keep Coordinated Universal Time in step with earth time, UT1. This is usually done only once a year and is intended to assist navigators.

LINE-10 - the tenth line out of 525 that make up a television picture. Line-10 has been adopted as a measurement point for time transfer.

LISSAJOUS PATTERNS - the patterns obtained on a cathode ray tube by driving both the horizontal and vertical deflection plates with particular signals. Lissajous patterns are used for frequency calibration and phase measurement.

LONG-TERM STABILITY - describes the frequency change of an oscillator that occurs with time. Long-term usually refers to changes over times longer than one second.

LOOP ANTENNA - an antenna used for radio reception at many different frequencies. It consists of a number of turns of wire usually in a shielded or pipe-type of enclosure. A loop antenna operates in the magnetic field of the signal and has a "figure-8" reception pattern which enables it to null out undesired signals.

MASER (MICROWAVE AMPLIFICATION BY SIMULATED EMISSION OF RADIATION) - a device that uses the natural oscillations of atoms and molecules to generate signals usually in the microwave region.

MAXIMUM USABLE FREQUENCY (MUF) - the highest frequency that can be used for radio transmissions without having the signal escape through the ionosphere. Signals below the maximum usable frequency will reflect off the ionosphere.

MEAN SOLAR TIME - time based on a mean solar day, which is the average length of all solar days in a solar year. The mean solar second is 1/86,400 of a mean solar day.

MIXER - an electrical circuit that mixes two signals and produces a new signal that is usually either the sum or the difference of the two inputs.

MODIFIED JULIAN DAY (MJD) - this is equal to the Julian date shifted so its origin occurs at midnight on November 17, 1858. That is, it differs from the Julian Day by 2,400,000.5 days exactly.

MULTIPATH - propagation of radio waves by more than one path from transmitter to receiver. This usually causes either fading or distortion.

NETWORK TELEVISION PROGRAM - any program that originates at either the East or West Coast networks of the U. S. broadcasters. It does not have to be a live program. It simply has to start out at the network studios.

NULL DETECTOR - an instrument that is used to detect the absence of a signal. Null detectors are usually more sensitive than peak detectors.

OFFSET (FREQUENCY) - frequency offset is the difference in the available frequency and the nominal frequency. It is often deliberate. In the case of the U. S. television networks at this writing, the offset is about -3000 parts in 1011. The term frequency offset is used interchangeably with the terms relative or fractional frequency.

OMEGA NAVIGATION SYSTEM - a low frequency (10 to 20 kHz) navigation system usually used in the hyperbolic mode.

OPTICAL DETECTION - detection of signal or its properties by optical means.

OVERTONE CRYSTALS - crystals designed to operate at a frequency higher than their fundamental. Crystal controlled oscillators as distinguished from crystal oscillators often use overtone crystals.

PERIOD MEASUREMENT - a measurement of frequency by measuring the duration of the periods in time units of the signal. Period measurements are always more accurate than frequency measurements unless the frequency being measured is as high as the frequency counter's time base. For example, a counter with a 1 MHz time base can be used in the period mode for greater accuracy for frequencies up to 1 MHz.

PHASE DETECTOR - a circuit or instrument that detects the difference in phase (sometimes in degrees or microseconds) between two signals. Phase detection is useful in high resolution frequency detection.

PHASE ERROR MULTIPLIERS - a scheme whereby the difference in phase between two signals is enhanced by multiplication and mixing techniques.

PHASE JUMP - the name given to sudden phase changes in a signal.

PHASE LOCK - a servo mechanism technique for causing one signal to follow another. They can be but do not have to be at the same frequency. Phase locking in terms of frequency sources is analogous to a mechanical servo where one shaft or wheel follows another.

PHASE SIGNATURE - a deliberate phase offset to identify a signal. WWVB as broadcast is deliberately phase shifted at 10 minutes after the hour. This aids the user by letting him know that he really is tracking WWVB and not some other signal.

POWER LINE FREQUENCY - In the United States the power lines are held to a nominal 60 hertz. Variations in this of about $\pm$ 50 ppm occur when the power system adjusts the frequency to regulate power flow.

PORTABLE CLOCK - any clock which is designed to be tranported. Usually it is an atomic clock with suitable batteries.

PRECISION - the reproducibility of a given measurement. For example, a clock pulse could be consistently measured to a precision of a few nanoseconds. Not to be confused with accuracy. The same clock that is measured to a precision of a few nanoseconds could be in error by several seconds.

PRESCALING - the process whereby a frequency out of range of a measurement instrument can be scaled down to within the range of that instrument.

PROPAGATION DELAY - the delay experienced by a signal as it traverses a radio path.

Q - the quality of a signal or detector or device. Q of coils is defined as the ratio of the inductive reactance to the resistance. It is also equal to the ratios of bandwidth of a resonator to its central frequency.

QUARTZ - a brilliant, crystalline mineral, mainly silicon dioxide.

RECEIVER DELAY - the delay experienced by the signal going from the antenna to the detector or output device, speaker, or oscilloscope terminal. Typical receiver delays in high frequency radio receivers are of the order of a few milliseconds.

REPRODUCIBILITY - the degree to which an oscillator will produce the same output frequency from one time to another after it has been properly aligned.

RESOLUTION - the degree to which a measurement can be determined is called the resolution of the measurement. A time interval counter might only resolve tenths of seconds or a few milliseconds.

RUBIDIUM (Rb) OSCILLATOR - an oscillator based on a resonance of rubidium gas.

SELECTIVITY - the ability of radio receiver circuits to separate one signal from another.

SENSITIVITY - the characteristic of a radio receiver that allows it to detect weak signals.

SHAPER - a circuit that provides uniform pulses from a variety of inputs. An example would be a circuit to change sine waves into sharp pulses.

SHORT-TERM STABILITY - a description of the frequency fluctuations caused by random noise in an oscillator. To properly state the short-term stability, the number of samples, averaging time, repetition time, and system bandwidth must be specified.

SIDEREAL TIME (SIGH DEER' EEE AL) - time based on observations of stars rather than the sun. Sidereal time is used by astronomers and a sidereal day is equal to about 23 hours, 56 minutes, and 4 seconds of solar time. Because it is based on observations of stars, it is more accurately determined than solar time.

SIGNAL-TO-NOISE RATIO (SNR) - the ratio of the strength of a radio signal to that of the noise. This is often a more useful term than simply signal strength.

SQUARING CIRCUIT - a circuit designed to produce square waves from an input waveform, quite often a sinusoid. This reduces the indecision about where zero crossings will occur.

STANDARD - a universally accepted reference. The NBS frequency standard is a cesium oscillator located at the National Bureau of Standards in Boulder, Colorado. Many frequencies generated and used in the United States are referenced to this standard.

STANDARD FREQUENCY BAND - radio bands allocated expressly for the purpose of distributing standard frequency and/or time. For example, all of the transmissions of WWV occur in standard frequency bands centered on 2.5, 5, 10, and 15 MHz.

STATE SELECTION - the process whereby atoms or molecules are separated, from atoms or molecules in a different state.

SUBCARRIER - an additional signal inserted along with the main carrier in radio or television transmission. The television color subcarrier is a signal added to the television format at about 3.58 MHz from the main carrier.

SUNSPOTS - dark spots sometimes seen on the surface of the sun which are related to magnetic and radio propagation disturbances on earth.

SUNSPOT CYCLES - cycles in the number of sunspots.

SYNCHRONIZATION - in the context of timing, synchronization means to bring two clocks or data streams into phase so they agree. They need not be on time.

TCXO (TEMPERATURE COMPENSATED CRYSTAL OSCIL-LATOR) - a crystal oscillator that contains special components to mini-mize the effect of temperature on the crystal frequency.

TICK - an electrical signal, usually a pulse, designed for use in resetting of dividers, etc. The WWV tick heard on its radio broadcast consists of 5 cycles of a 1 kHz sine wave.

TIME BASE - in a frequency counter that oscil-lator which provides the timing signals for measurement and control. The accuracy of the counter is directly related to the time base.

TIME INTERVAL COUNTER - a counter designed to measure the time between events, zero crossings for example.

TIME OF COINCIDENCE (LORAN) - that time when the Loran signal is coincident with an exact second of Coordinated Universal Time. These times of coincidence will occur about every 10 or 15 minutes and have been tabulated by the U. S. Naval Observ-atory.

TRANSFER STANDARDS - the name given to a sig-nal that is used to perform a cali-bration when it can be referenced to a standard. An example would be the use of a VLF frequency source to transfer frequency to a location based on its calibration at NBS.

TRIGGER ERROR - The error associated with false triggering caused by phase noise or jitter on a signal.

TTL INTEGRATED CIRCUITS - a family of integra-ted circuits that uses transistor-transistor logic. At this writing TTL has become the most popular family of integrated circuits and is characterized by fast rise times. Unless deliberately modified, output levels are nominally zero to 5 volts positive.

UNIVERSAL TIME (UT) FAMILY - Universal Time is given in several ways. Apparent solar time is first corrected by the equation of time to mean solar time, UT0. It is then again corrected for migration of the earth's poles to obtain UT1. This is further cor-rected for periodicity of unknown origin to obtain UT2.

VFO - variable frequency oscillator as dis-tinguished from a fixed frequency oscillator such as a crystal.

WAVE ANGLE - the angle above the surface of the earth at which a signal is transmitted from or received by an antenna.

WAVEMETER - a meter for measuring frequency. Wavemeters usually work by having a coil that collects part of the energy from a transmitter or a transmission line and displays resonance on a meter movement.

ZERO BEAT - that condition between two signals when no beat is heard or seen be-tween the signals; that is, the frequencies are equal.

227

234

This manual has been written for the person who needs information on making time and frequency measurements. It has been written at a level that will satisfy those with a casual interest as well as laboratory engineers and technicians who use time and frequency every day. It gives a brief history of time and frequency, discusses the roles of the National Bureau of Standards, the U. S. Naval Observatory, and the International Time Bureau, and explains how time and frequency are internationally coordinated. It also explains what time and frequency services are available and how to use them. It discusses the accuracies that can be achieved using the different services as well as the pros and cons of using various calibration methods.

Waste Heat Management Guidebook

A typical plant can save about 20 percent of its fuel—just by installing waste heat recovery equipment. But with so much equipment on the market, how do you decide what's right for you?

Find the answers to your problems in the *Waste Heat Management Guidebook*, a new handbook from the Commerce Department's National Bureau of Standards and the Federal Energy Administration.

The *Waste Heat Management Guidebook* is designed to help you, the cost-conscious engineer or manager, learn how to capture and recycle heat that is normally lost to the environment during industrial and commercial processes.

The heart of the guidebook is 14 case studies of companies that have recently installed waste heat recovery systems and profited. One of these applications may be right for you, but even if it doesn't fit exactly, you'll find helpful approaches to solving many waste heat recovery problems.

In addition to case studies, the guidebook contain information on:

* sources and uses of waste heat
* determining waste heat requirements
* economics of waste heat recovery
* commercial options in waste heat recovery equipment
* instrumentation
* engineering data for waste heat recovery
* assistance for designing and installing waste heat systems

To order your copy of the *Waste Heat Management Guidebook*, send $2.75 per copy (check or money order) to Superintendent of Documents, U.S. Government Printing Office, Washington, D.C. 20402. A discount of 25 percent is given on orders of 100 copies or more mailed to one address.

The *Waste Heat Management Guidebook* is part of the EPIC industrial energy management program aimed at helping industry and commerce adjust to the increased cost and shortage of energy.

U.S. DEPARTMENT OF COMMERCE/National Bureau of Standards
FEDERAL ENERGY ADMINISTRATION/Energy Conservation and Environment

There's a new look to...

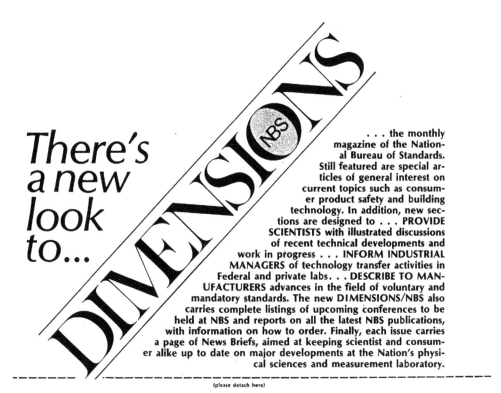

... the monthly magazine of the National Bureau of Standards. Still featured are special articles of general interest on current topics such as consumer product safety and building technology. In addition, new sections are designed to ... PROVIDE SCIENTISTS with illustrated discussions of recent technical developments and work in progress ... INFORM INDUSTRIAL MANAGERS of technology transfer activities in Federal and private labs. . . DESCRIBE TO MANUFACTURERS advances in the field of voluntary and mandatory standards. The new DIMENSIONS/NBS also carries complete listings of upcoming conferences to be held at NBS and reports on all the latest NBS publications, with information on how to order. Finally, each issue carries a page of News Briefs, aimed at keeping scientist and consumer alike up to date on major developments at the Nation's physical sciences and measurement laboratory.

(please detach here)

SUBSCRIPTION ORDER FORM

Enter my Subscription To DIMENSIONS/NBS at $11.00. Add $2.75 for foreign mailing. No additional postage is required for mailing within the United States or its possessions. Domestic remittances should be made either by postal money order, express money order, or check. Foreign remittances should be made either by international money order, draft on an American bank, or by UNESCO coupons.

☐ Remittance Enclosed (Make checks payable to Superintendent of Documents)

☐ Charge to my Deposit Account No.

Send Subscription to:

NAME-FIRST, LAST

COMPANY NAME OR ADDITIONAL ADDRESS LINE

STREET ADDRESS

CITY   STATE   ZIP CODE

PLEASE PRINT

MAIL ORDER FORM TO:
Superintendent of Documents
Government Printing Office
Washington, D.C. 20402

NBS TECHNICAL PUBLICATIONS

PERIODICALS

JOURNAL OF RESEARCH—The Journal of Research of the National Bureau of Standards reports NBS research and development in those disciplines of the physical and engineering sciences in which the Bureau is active. These include physics, chemistry, engineering, mathematics, and computer sciences. Papers cover a broad range of subjects, with major emphasis on measurement methodology and the basic technology underlying standardization. Also included from time to time are survey articles on topics closely related to the Bureau's technical and scientific programs. As a special service to subscribers each issue contains complete citations to all recent Bureau publications in both NBS and non-NBS media. Issued six times a year. Annual subscription: domestic $17; foreign $21.25. Single copy, $3 domestic; $3.75 foreign.

NOTE: The Journal was formerly published in two sections: Section A "Physics and Chemistry" and Section B "Mathematical Sciences."

DIMENSIONS/NBS—This monthly magazine is published to inform scientists, engineers, business and industry leaders, teachers, students, and consumers of the latest advances in science and technology, with primary emphasis on work at NBS. The magazine highlights and reviews such issues as energy research, fire protection, building technology, metric conversion, pollution abatement, health and safety, and consumer product performance. In addition, it reports the results of Bureau programs in measurement standards and techniques, properties of matter and materials, engineering standards and services, instrumentation, and automatic data processing. Annual subscription: domestic $11; foreign $13.75.

NONPERIODICALS

Monographs—Major contributions to the technical literature on various subjects related to the Bureau's scientific and technical activities.

Handbooks—Recommended codes of engineering and industrial practice (including safety codes) developed in cooperation with interested industries, professional organizations, and regulatory bodies.

Special Publications—Include proceedings of conferences sponsored by NBS, NBS annual reports, and other special publications appropriate to this grouping such as wall charts, pocket cards, and bibliographies.

Applied Mathematics Series—Mathematical tables, manuals, and studies of special interest to physicists, engineers, chemists, biologists, mathematicians, computer programmers, and others engaged in scientific and technical work.

National Standard Reference Data Series—Provides quantitative data on the physical and chemical properties of materials, compiled from the world's literature and critically evaluated. Developed under a worldwide program coordinated by NBS under the authority of the National Standard Data Act (Public Law 90-396).

NOTE: The principal publication outlet for the foregoing data is the Journal of Physical and Chemical Reference Data (JPCRD) published quarterly for NBS by the American Chemical Society (ACS) and the American Institute of Physics (AIP). Subscriptions, reprints, and supplements available from ACS, 1155 Sixteenth St., NW, Washington, DC 20056.

Building Science Series—Disseminates technical information developed at the Bureau on building materials, components, systems, and whole structures. The series presents research results, test methods, and performance criteria related to the structural and environmental functions and the durability and safety characteristics of building elements and systems.

Technical Notes—Studies or reports which are complete in themselves but restrictive in their treatment of a subject. Analogous to monographs but not so comprehensive in scope or definitive in treatment of the subject area. Often serve as a vehicle for final reports of work performed at NBS under the sponsorship of other government agencies.

Voluntary Product Standards—Developed under procedures published by the Department of Commerce in Part 10, Title 15, of the Code of Federal Regulations. The standards establish nationally recognized requirements for products, and provide all concerned interests with a basis for common understanding of the characteristics of the products. NBS administers this program as a supplement to the activities of the private sector standardizing organizations.

Consumer Information Series—Practical information, based on NBS research and experience, covering areas of interest to the consumer. Easily understandable language and illustrations provide useful background knowledge for shopping in today's technological marketplace.

Order the above NBS publications from: Superintendent of Documents, Government Printing Office, Washington, DC 20402.

Order the following NBS publications—FIPS and NBSIR's—from the National Technical Information Services, Springfield, VA 22161.

Federal Information Processing Standards Publications (FIPS PUB)—Publications in this series collectively constitute the Federal Information Processing Standards Register. The Register serves as the official source of information in the Federal Government regarding standards issued by NBS pursuant to the Federal Property and Administrative Services Act of 1949 as amended, Public Law 89-306 (79 Stat. 1127), and as implemented by Executive Order 11717 (38 FR 12315, dated May 11, 1973) and Part 6 of Title 15 CFR (Code of Federal Regulations).

NBS Interagency Reports (NBSIR)—A special series of interim or final reports on work performed by NBS for outside sponsors (both government and non-government). In general, initial distribution is handled by the sponsor; public distribution is by the National Technical Information Services, Springfield, VA 22161, in paper copy or microfiche form.

BIBLIOGRAPHIC SUBSCRIPTION SERVICES

The following current-awareness and literature-survey bibliographies are issued periodically by the Bureau:

Cryogenic Data Center Current Awareness Service. A literature survey issued biweekly. Annual subscription: domestic $25; foreign $30.

Liquefied Natural Gas. A literature survey issued quarterly. Annual subscription: $20.

Superconducting Devices and Materials. A literature survey issued quarterly. Annual subscription: $30. Please send subscription orders and remittances for the preceding bibliographic services to the National Bureau of Standards, Cryogenic Data Center (736) Boulder, CO 80303.

U.S. DEPARTMENT OF COMMERCE
National Bureau of Standards
Washington, D.C. 20234

OFFICIAL BUSINESS

Penalty for Private Use, $300

SPECIAL FOURTH-CLASS RATE
BOOK

1298